Continuous-Flow Chemistry in the Research Laboratory

Toma Glasnov

Continuous-Flow Chemistry in the Research Laboratory

Modern Organic Chemistry in Dedicated Reactors at the Dawn of the 21st Century

Springer

Toma Glasnov
Institute of Chemistry
University of Graz
Graz
Austria

ISBN 978-3-319-32194-3 ISBN 978-3-319-32196-7 (eBook)
DOI 10.1007/978-3-319-32196-7

Library of Congress Control Number: 2016940841

Printed on acid-free paper

This Springer imprint is published by Springer Nature
The registered company is Springer International Publishing AG Switzerland

Preface

As a result of new environment regulations, safety concerns, and the economical situation after the last crisis in 2008, there is a strong need of new innovative, environmentally friendly synthetic routes and enabling technologies to meet the new requirements. In the last few years, we have witnessed a steady growth in the field of continuous flow synthesis. The rising interest in this technology is in a direct relation with the recognition that this technique actually provides various advantages, especially in dealing with potentially hazardous chemistries, handling thermal runaways, or efficient mixing requirements. Despite the industrial background, continuous flow processing has slowly breached the barrier to academia and is now often considered as the logical choice for scaling up laboratory syntheses. However, as with every new technology, the obstacle of missing information and education on the basic principles, common problems, already existing protocols, and applications prevents its implementation in the daily research. Thus, the aim of this book is to give the reader a structured overview of known synthetic procedures involving the use of dedicated continuous flow instrumentation published during the last 15 years—the dawn of the twenty-first century. Although there are a large number of papers dealing with continuous flow processing (engineering, theoretical background, modelling, etc.), only those references dealing with organic synthesis examples are incorporated. Nevertheless, I would like to extend my apologies to all the scientists whose research findings could not be cited or discussed here.

Finally, I would like to acknowledge Dr. David Obermayer and Dr. Bernadett Bacsa for the help, discussions, and suggested improvements on the manuscript.

Graz, Austria Toma Glasnov

Contents

Chapter 1
Continuous Flow Synthesis: A Short Perspective

In the past few years, continuous flow processing has slowly started to find place in academic research. Considered more of an industrial value for large-scale synthesis in chemical industry, it took more than half a century for academia to slowly adopt this technology for small-scale laboratory synthesis. Although there are clear benefits, especially whenever working with hazardous intermediates, that have to be generated in situ, or rapid heat dissipation and efficient mixing are needed, the general use of continuous flow synthesis on a daily basis in the modern research laboratory remains controversial. Still, flow synthesis appears to be seen as a curiosity and merely an expert tool among the many other and more "traditional" synthesis techniques. As such, the plethora of recent examples found in the literature remains focused on exploring the capabilities of the available equipment for optimizing already established syntheses and rarely a novelty from a chemical point of view is found. The challenge of processing heterogeneous reactions and reagents, highly viscous or highly corrosives materials, as well as the required time and labor investment for developing a running flow process further hurdles. Nevertheless, and in many instances, the use of dedicated flow equipment has proven its value and can bring undisputable advantage for the synthetic chemist in the research laboratory—continuous flow hydrogenation, ozonolysis, or lithium exchange reactions are just some of these synthetic examples. Although continuous flow technology offers a technically unique way to perform synthetic reactions, the question of whether to use this technique for a chemical transformation should be taken by an experienced chemist.

Very similar to the boom of microwave-assisted synthesis over the first decade of the twenty-first century and the remarkable improvements it brought to academia and industry research by tremendously increasing the daily output of a research laboratory, continuous flow processing is seen as the next "hot topic" in synthetic technology. Although the first reports of continuous flow experimentation from a research laboratory date back in the middle and late twentieth century, it was only after the turn of the twenty-first century that a slow growth in the area could be seen. The lack of dedicated equipment, the insufficient methodological knowledge, and

T. Glasnov, *Continuous-Flow Chemistry in the Research Laboratory*,
DOI 10.1007/978-3-319-32196-7_1

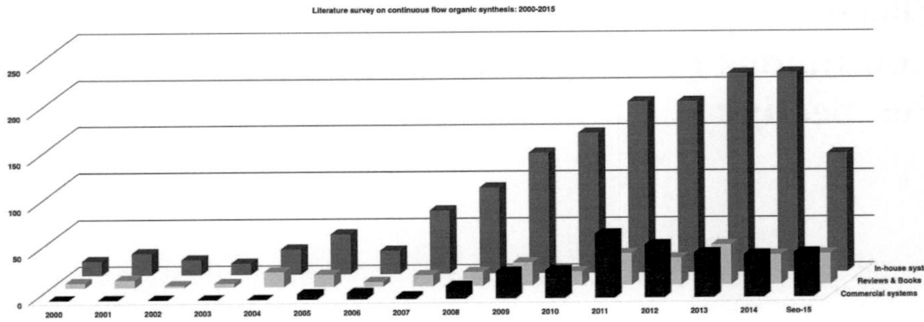

Fig. 1.1 Publications on continuous flow organic synthesis (2000–September 2015). Only articles dealing with synthetic organic chemistry examples were included

the absence of an educational link between chemical engineering and synthetic organic chemistry have been responsible for the slow uptake flow synthetic techniques by the scientific community. Interestingly, the exact same problems have been encountered previously with the introduction of microwave processing in synthetic chemistry. Through the work of only few research groups located predominately at university campuses in the USA, UK, and Japan, the scientific community started to get slowly aware of the new technology. However, it was only after the introduction of the first few dedicated flow instruments on the market that certain interest among researchers around the world started to develop. Thus, in the last 10–15 years, a rapidly increasing number of publications exploiting this new technology in all areas of organic synthesis have been published (see Fig. 1.1).

Although this technique will probably by far not reach the acceptance of microwave synthesis, it is presently enjoying high popularity. Today, an assortment of several books [1–11], special issues of synthetic chemistry journals, and an extensive number of review articles [12–68] cover the published literature from various viewpoints.

This book emphasizes on selected examples of continuous flow processing in organic synthesis from the last decade—2005 until September 2015. A considerable number of published work has already covered the basics of continuous flow processing with extensive information on processing techniques, as well as the design and manufacture of "build-it-yourself" continuous flow devices. Thus, the focus in this book is set on highlighting synthetic applications in dedicated commercially available continuous flow systems assuring adequate reproducibility of optimized protocols in any scientific laboratory. Continuous microwave protocols are not part of this overview. In terms of processing techniques, various approaches are discussed—heterogeneous and homogenous reactions, single and multiple step syntheses, and processes at various temperature regimes and pressures. Among the ca. 1900 original publications published over the covered time period, a simple analysis shows that in ca. 23 % of the published work, dedicated continuous flow equipment has been employed (Fig. 1.1). Additionally, it also reveals that the current trend among scientists still favors the use of in-house build devices and systems. Another focus of the following overview is on continuous flow examples of interest to organic/medicinal chemists working in research laboratories in industry or academia. The large amount of publications dictates the information in this book to be arranged as a mix of graphical and text format, discussing shortly the presented chemistry examples. This book is therefore primarily intended as a resource of ideas and references for a wide audience of organic chemists.

References

Selected Recent Books on Organic Synthesis Under Continuous Flow Conditions

1. V. Hessel, S. Hardt, H. Löwe, *Chemical Micro Process Engineering* (Wiley-VCH, Weinhem, 2004)
2. J. Yoshida, *Flash Chemistry: Fast Organic Synthesis in Microsystems* (Wiley-Blackwell, Chichester, 2008)
3. T. Dietrich, *Principles and Applications of Chemical Microreactors* (Wiley, Hoboken, 2008)
4. T. Wirth (ed.), *Microreactors in Organic Synthesis and Catalysis* (Wiley-VCH, Weinheim, 2008)
5. V. Hessel, A. Renken, J.C. Schouten, J. Yoshida (eds.), *Micro Process Engineering. A Comprehensive Handbook,* vols. 1–3 (Wiley-VCH, Weinheim, 2009)
6. C. Wiles, P. Watts, *Micro Reaction Technology in Organic Synthesis* (CRC Press, Boca Raton, 2011)
7. W. Reschetilowski (ed.), *Microreactors in Preparative Chemistry* (Wiley-VCH, Weinheim, 2013)
8. T. Wirth (ed.), *Microreactors in Organic Synthesis and Catalysis. Second, Completely Revised and Enlarged Edition* (Wiley-VCH, Weinheim, 2013)
9. F. Darvas, G. Dorman, V. Hessel (eds.), *Flow Chemistry,* vols. 1-2 (De Gruyter GmbH, Berlin/ Boston, 2014)
10. J. Yoshida, *Basics of Flow Microreactor Synthesis* (Springer, Tokyo, 2015)
11. V. Hessel, D. Kralissh, N. Kockmann, *Novel Process Windows—Innovative Gates to intensified and Sustainable Chemical Processes* (Wiley-VCH, Weinheim, 2015)

Selected Recent Reviews on Organic Synthesis Under Continuous Flow Conditions (2010–2015)

12. C.G. Frost, L. Mutton, Green Chem. **12**, 1687 (2010)
13. T. Illg, P. Lob, V. Hessel, Bioorg. Med. Chem. **18**, 3707 (2010)
14. S.V. Ley, Tetrahedron **66**, 6270 (2010)
15. S. Marre, K.F. Jensen, Chem. Soc. Rev. **39**, 1183 (2010)

16. J.P. McMullen, K.F. Jensen, Annu. Rev. Anal. Chem. **3**, 19 (2010)
17. T. Razzaq, C.O. Kappe, Chem. Asian. J. **5**, 1274 (2010)
18. F.E. Valera, M. Quaranta, A. Moran, J. Blacker, A. Armstrong, J.T. Cabral, D.G. Blackmond, Angew. Chem. Int. Ed. **49**, 2478 (2010)
19. D. Webb, T.F. Jamison, Chem. Sci. **1**, 675 (2010)
20. J. Yoshida, Chem. Rec. **10**, 332 (2010)
21. C. Wiles, P. Watts, Adv. Chem. Eng. **38**, 103 (2010)
22. A. Cukalovic, J.-C.M.R. Monbaliu, C. Stevens, Top. Heterocycl. Chem. **23**, 161 (2010)
23. T.N. Glasnov, C.O. Kappe, J. Heterocycl. Chem. **48**, 11 (2011)
24. R.L. Hartman, J.P. McMullen, K.F. Jensen, Angew. Chem. Int. Ed. **50**, 7502 (2011)
25. M. Rasheed, T. Wirth, Angew. Chem. Int. Ed. **50**, 357 (2011)
26. J. Wegner, S. Ceylan, A. Kirschning, Chem. Commun. **47**, 4583 (2011)
27. C. Wiles, P. Watts, Chem. Commun. **47**, 6512 (2011)
28. J. Yoshida, H. Kim, A. Nagaki, ChemSusChem **4**, 331 (2011)
29. T.N. Glasnov, C.O. Kappe, Chem. Eur. J. **17**, 11956 (2011)
30. M. Irfan, T.N. Glasnov, C.O. Kappe, ChemSusChem **4**, 300 (2011)
31. T. Nöel, S.L. Buchwald, Chem. Soc. Rev. **40**, 5050 (2011)
32. J.W. Tucker, Y. Zhang, T.F. Jamison, C.R.J. Stephenson, Angew. Chem. Int. Ed. **51**, 4144 (2012)
33. A. Kirschning, L. Kupracz, J. Hartwig, Chem. Lett. **41**, 562 (2012)
34. J. Wegner, S. Ceylan, A. Kirschning, Adv. Synth. Catal. **354**, 17 (2012)
35. C. Wiles, P. Watts, Green Chem. **14**, 38 (2012)
36. L. Malet-Sanz, F. Susanne, J. Med. Chem. **55**, 4062 (2012)
37. M. Oelgemoeller, Chem. Eng. Technol. **35**, 1144 (2012)
38. T. Chinnusamy, S. Yudha, S.M. Hager, P. Kreitmeier, O. Reiser, ChemSusChem **5**, 247 (2012)
39. C.B. McPake, G. Sandford, Org. Process Res. Dev. **16**, 844 (2012)
40. T. Tsubogo, T. Ishiwata, S. Kobayashi, Angew. Chem. Int. Ed. **52**, 6590 (2013)
41. T. Nöel, V. Hessel, ChemSusChem **6**, 405 (2013)
42. I.R. Baxendale, J. Chem. Technol. Biotechnol. **88**, 519 (2013)
43. D.T. McQuade, P.H. Seeberger, J. Org. Chem. **78**, 6384 (2013)
44. J.C. Pastre, D.L. Browne, S.V. Ley, Chem. Soc. Rev. **42**, 8849 (2013)
45. J. Yoshida, A. Nagaki, D. Yamada, Drug Discov. Today Technol. **10**, e53 (2013)
46. J. Yoshida, Y. Takahashi, A. Nagaki, Chem. Commun. **49**, 9896 (2013)
47. V. Hessel, D. Kralish, N. Kockmann, T. Nöel, Q. Wang, ChemSusChem **6**, 746 (2013)
48. S.G. Newman, K.F. Jensen, Green Chem. **15**, 1456 (2013)
49. A. Puglisis, M. Benaglia, V. Chiroli, Green Chem. **15**, 1790 (2013)
50. T. Rodrigues, P. Schneider, G. Schneider, Angew. Chem. Int. Ed. **53**, 5750 (2014)
51. K.S. Elvira, X. Casadevall i Solvas, R.C.R. Wootton, A. J. deMello, Nat. Chem. **5**, 905 (2013)
52. T. Fukuyama, T. Totoki, I. Ryu, Green Chem. **16**, 2042 (2014)
53. Y. Su, N.J.W. Straathof, V. Hessel, T. Nöel, Chem. Eur. J. **20**, 10562 (2014)
54. C. Wiles, P. Watts, Green Chem. **16**, 55 (2014)
55. T. Fukuyama, T. Totoki, I. Ryu, Green Chem. **16**, 2042 (2014)
56. L. Vaccaro, D. Lanari, A. Marrocchi, G. Strappaveccia, Green Chem. **16**, 3680 (2014)
57. S. Fuse, Y. Mifune, N. Tanabe, T. Takahashi, Synlett **25**, 2087 (2014)
58. K. Gilmore, P. Seeberger, Chem. Rec. **14**, 410 (2014)
59. K. Hargrove, G. Jones, Curr. Radiopharm. **7**, 36 (2014)
60. B. Gutmann, D. Cantillo, C.O. Kappe, Angew. Chem. Int. Ed. **54**, 6688 (2015)
61. S.V. Ley, D.E. Fitzpatrick, R.M. Myers, C. Battilocchio, R.J. Ingham, Angew. Chem. Int. Ed. **54**, 10122 (2015)
62. M. Baumann, I.R. Baxendale, Beilstein J. Org. Chem. **11**, 1194 (2015)
63. J. Bao, G.K. Tranmer, Chem. Commun. **51**, 3037 (2015)

64. F.G. Finelli, L.S.M. Miranda, R.O.M.A. de Souza, Chem. Commun. **51**, 3708 (2015)
65. B.J. Deadman, S.G. Collins, A.R. Maguire, Chem. Eur. J. **21**, 2298 (2015)
66. S.T.R. Müller, T. Wirth, ChemSusChem **8**, 245 (2015)
67. P.J. Cossar, L. Hizartzidis, M.I. Simone, A. McCluskey, C.P. Gordon, Org. Biomol. Chem. **13**, 7119 (2015)
68. H.P.L. Gemoets, Y. Su, M. Shang, V. Hessel, R. Luque, T. Nöel, Chem. Soc. Rev. **45**, 83 (2016)

Chapter 2
Equipment Overview

2.1 The "Build-It-Yourself (BIY)" Approach

Due to the still relatively high costs of dedicated commercial flow instrumentation, the majority of chemists around the world practice the "build-it-yourself" (BIY) or also "do-it-yourself" (DIY) approach. Ever since the few early reports on flow synthesis [1, 2], the most preferred option in the community to date is to assemble continuous flow devices for synthetic purposes using redundant parts from HPLC and GC instrumentation. However, this approach is often associated with major reproducibility issues. Indeed, the published flow procedures from the last decade have been only very rarely reproduced and further employed beyond the original reports. For this reason, the research performed in such devices remains beyond the scope of this book.

2.2 Dedicated Continuous Flow Systems for Organic Synthesis

With growing interest in continuous flow synthesis on laboratory scale, the demand for sophisticated instrumentation has also increased in recent years. With the market introduction of the modular AFRICA™ system by Syrris, the H-Cube™ flow hydrogenator by ThalesNano, the Ehrfeld module platform for flow, the CPC-College system, and few other platforms between 2005 and 2006, automated continuous flow synthesis became available for laboratory-scale synthesis. Safe and reproducible work is now possible without any "engineering" efforts for "BIY" flow systems. The major players on the market for laboratory continuous flow equipment with the various instrumental equipments are summarized below.

T. Glasnov, *Continuous-Flow Chemistry in the Research Laboratory*,
DOI 10.1007/978-3-319-32196-7_2

2.2.1 *ThalesNano Nantechnology Inc. [3, 4]*

2.2.1.1 H-Cube^{TM}

The H-Cube^{TM} was introduced as a stand-alone flow hydrogenation reactor back in 2005. The bench-top, shoebox-sized system easily fits into any laboratory fume hood and allows straightforward access to hydrogen-involving reactions under flow conditions. Eliminating the need of a specially equipped hydrogenation room and handling of pressurized hydrogen gas bottles, the instrument allows fast, safe, and cost-efficient processing by on-demand hydrogen generation through water electrolysis.

A piston pump delivers the substrate–solvent mixture into the system where it is mixed with the generated hydrogen gas, before passing over a cartridge packed with a heterogeneous catalyst. The reaction mixture can be heated up to 100 °C and pressurized up to 100 bar. The instrument can be run in three different modes—"no hydrogen mode," "full mode," and "controlled mode." The "no hydrogen" mode allows the use of the instrument for different chemistries besides hydrogenation. In "full mode," all the generated hydrogen is mixed with the reaction mixture at atmospheric pressure, while the "controlled mode" allows pressurizing the system with selected amount of hydrogen up to 100 bar. The flow rate of the reaction mixture can be selected in the range 0.5–3 mL/min via the touchscreen and is communicated to the external pump. The reaction takes place in the heated cartridge holder. The cartridge concept (CatCart®) allows the use of various commercial solid catalysts as well as newly developed ones. Three different sizes of stainless steel CatCarts® are available —30, 50, and 70 mm in length. The use of CatCarts® eliminates the need of catalyst removal after the reaction has finished. With the H-Cube^{TM} hydrogenator, amounts in the range of several milligrams up to 10 g can be processed successfully.

2.2.1.2 H-Cube Pro^{TM}

The H-Cube Pro^{TM} (Fig. 2.1) is a newer generation of the H-Cube family, integrating the features of previous systems [6–9] while giving the opportunity to widen the reaction scope that can be explored under flow conditions. New features include:

- Two hydrogen cells to generate up to 60 mL/min hydrogen
- Reaction temperatures in the range of 10–150 °C
- Support of external modules—gas module for the controlled supply of gases other than hydrogen, Phoenix Flow Reactor (see below), etc.
- Full automation and external software control

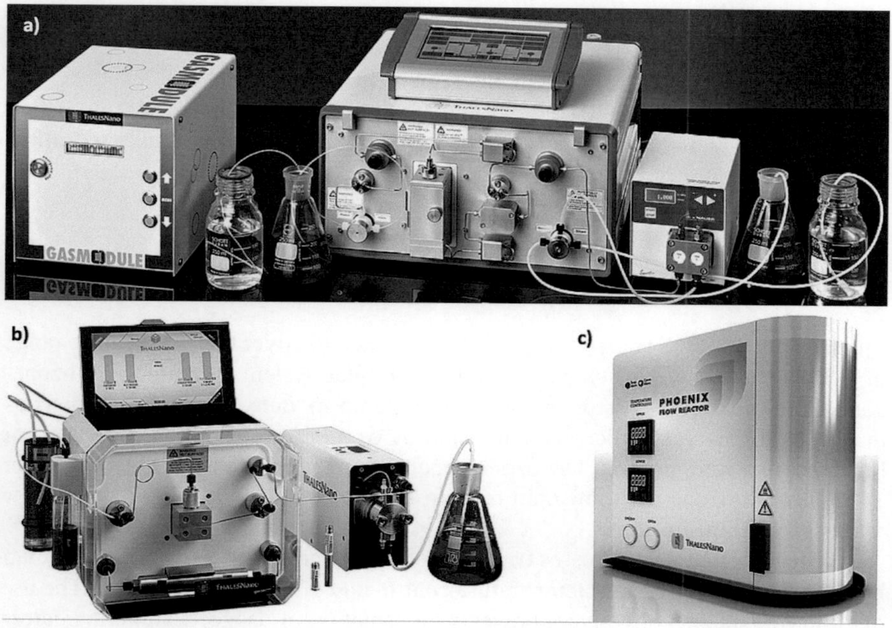

Fig. 2.1 ThalesNano instruments—(**a**) H-Cube Pro™; (**b**) H-Cube Mini™; (**c**) Phoenix™ reactor

2.2.1.3 H-Cube Midi™

The H-Cube Midi™ is developed as scale-up version of the H-Cube concept. In this manner, this flow hydrogenator is able to deliver an increased productivity of up to 500 g/day of product. The reaction mixture can be flowed through the system with an automatically controlled piston pump at flow rates of 3–25 mL/min. The working reaction temperature can be up to 150 °C, and CatCarts® of 9.5 × 90 mm in size are used, able to carry several grams of catalyst.

2.2.1.4 H-Cube Mini™

The H-Cube Mini™ (Fig. 2.1) is developed for education purposes in academia and represents a simplified version of the H-Cube instrument.

2.2.1.5 Phoenix Flow Reactor™

The Phoenix Flow Reactor™ (Fig. 2.1) is a high-temperature reactor for heterogeneous or homogenous reaction in flow conditions. It combines the properties of two earlier instruments—the X-Flash and the X-Cube [5–7]. It can work as an add-on for the H-Cube and H-Cube Pro reactors or a as a stand-alone instrument. The reactor works with capillary tubing (coils) from stainless steel, Hastelloy®, or Teflon®. Respectively, reaction temperatures in the range of 150–450 °C are accessible. The standard 30 and 70 mm CatCarts® can be used in temperature regimes up to 250 °C. Specially developed 125 and 250 mm CatCarts® allow working conditions of up to 450 °C (petrochemical applications).

2.2.1.6 IceCube™ Flow Reactor

The IceCube Flow Reactor™ (Fig. 2.2) is designed to cover the temperature range of −70–80 °C. It is a software-controlled, modular system containing an ozone-generating module, a reactor module, and a pump module. It enables the performance of highly energetic reactions such as ozonolysis, azidation, nitration, or lithiation in a safe manner. The ozone generator (ozone module) is able to deliver 14 % (v/v) of ozone at 20 mL/min oxygen flow rate. The applicable oxygen flow rate is 10–100 mL/min.

The reactor module possesses two reaction plates, equipped with Peltier cooling/heating modules for precise temperature control and a Teflon reaction line. The use of an in-line quench effectively prevents the isolation of dangerous intermediates.

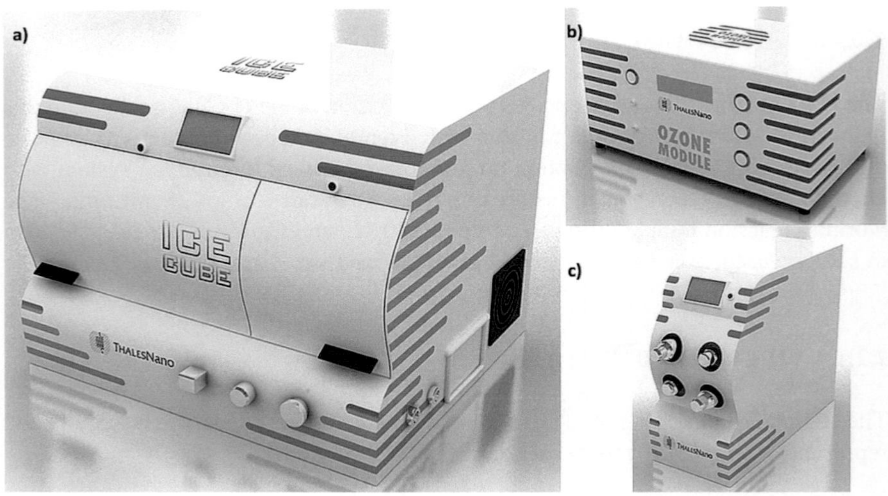

Fig. 2.2 ThalesNano instruments—(**a**) IceCube reactor; (**b**) ozone module; (**c**) pump module

2.2.2 *Syrris [8]*

2.2.2.1 ASIA Modular System™

The ASIA modular system (Fig. 2.3) allows a wide range of configuration options to meet the synthetic requirements of various chemical processes. The flow system can be controlled either "manually" or interfaced to a computer. The specifications of the system include the following:

- Temperature regimes: −15 to + 250 °C
- Liquid phase reactor volumes: 62.5 μL, 250 μL, 1 mL, 4 mL, and 16 mL
- Solid phase reactor volumes: 0.7, 2.4, 5.6, and 12 mL
- Working pressures: 0–20 bar
- Flow rates: 1 μL/min–10 mL/min using continuous syringe pumps
- Wetted materials: glass, Teflon®, PCTFE, stainless steel, and Hastelloy®

The system allows the implementation of tube (coil), chip, and glass column reactors as well as the realization of multistep syntheses, where the reactors can be combined and used sequentially. An interesting module is the FLLEX™ liquid–liquid extractor, allowing an in-line extraction as integrated purification step.

2.2.2.2 AFRICA Modular System™

The AFRICA system is a highly sophisticated, fully automated, modular flow system for R&D chemists enabling the production of kilogram quantities of product overnight [9].

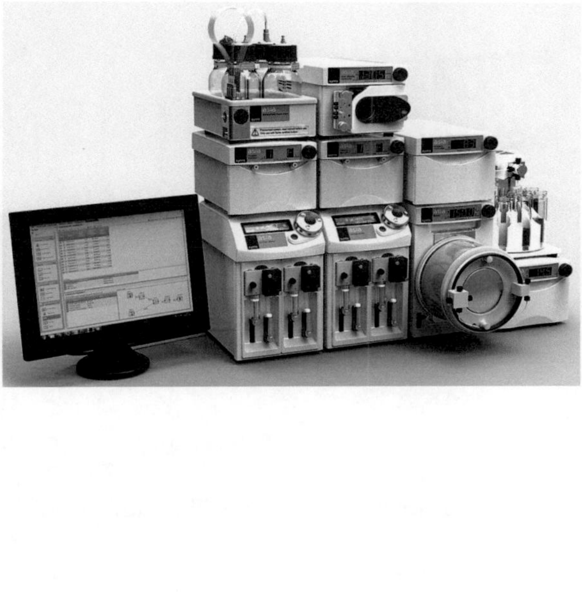

Fig. 2.3 SYRRIS System Asia™ 230 modular system

2.2.3 *Vapourtec Ltd [10]*

2.2.3.1 R-Series Modular System

The R-Series modular system (Fig. 2.4) consists of two main modules with different reactor options—tube (coil), column, tube in tube, etc. The R-series pump module (R1 or R2 in various configurations) allows working with flow rates of 0.05–50 mL/min and 10–200 bar pressure. An acid-resistant modification of the pump is also available which allows the use of concentrated sulfuric and fuming nitric acid. The R-series reactor module (R4 module) provides four independently temperature-controlled reactor positions for using exchangeable reactors:

- Standard PFA coiled tube reactor: 2, 5, and 10 mL reactor volumes
- Stainless steel 316 or Hastelloy coiled tube reactor: 2, 5, and 10 mL reactor volumes; usable for reactions up to 250 °C
- Cooled coil reactor: for reactions at −70 °C to ambient
- Glass column reactor: −40 to 150 °C temperature regimes covered; for solid reagents/catalysts/scavengers

The fully automated version features software control, fraction collector, additional pump line, and an autosampler.

2.2.3.2 E-Series Modular System

The E-series modular system (Fig. 2.4) is a newer development from Vapourtec. It is available in four basic configurations—easy scholar, easy polymer, easy medchem, and easy photochem. All of the configurations have three V3 model

a) b)

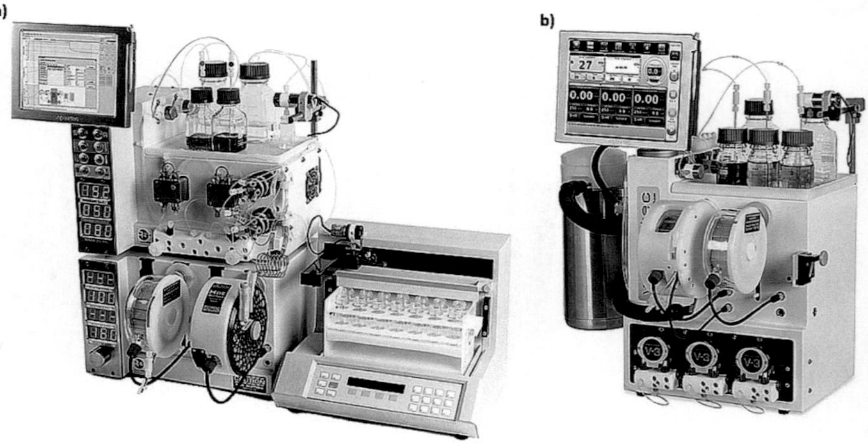

Fig. 2.4 Vapourtec Instruments—(**a**) R-series; (**b**) E-series

pumps, able to handle light suspensions and even slurries. The four flow systems support up to two reactor positions which can accommodate the full range of reactors available as separate modules from Vapourtec. The easy photochem system is intended for photochemical syntheses and can be equipped with either a LED light source (365–500 nm) or with a high-intensity, medium-pressure Hg lamp combined with a plethora of optical filters for isolated irradiation wavelengths. Additional chemistry tools are integrated into the software package of all models.

2.2.4 Uniqsis Ltd [11]

2.2.4.1 FlowSyn™

The FlowSyn is a compact flow system (Fig. 2.5) with two integrated high-pressure pumps (up to 20 mL/min flow rate, up to 200 bar pressure) and two independent heated reactor modules—for column or chip reactors (up to 150 °C) and a coil reactor heater (up to 260 °C). Combining a chip reactor as a mixing device with a coil reactor is possible. The coil reactors are available in various materials—stainless steel, Hastelloy, copper, PTFE, and PFA. Glass columns and static mixers/reactor chips are also available on demand. On a modular basis, different add-on devices can be used—a fraction collector (Multi-X), liquid handler (Auto-LF), additional pump (Binary Pumping Module), or heater/chiller module (−88 °C Polar Bear™; −40 to 150 °C, Polar Bear Plus™)—and a higher throughput version (up to 100 mL/min; Maxi) etc.

2.2.4.2 FlowStart™

The FlowStart™ system is a modular entry level system (Fig. 2.5). It is a combination of two high-pressure pumps and a HotCoil™ reactor station, both controlled by the FlowStart™ software via LAN. The flow rate can be adjusted between 0.01–20 mL/min at pressures of up to 100 bar. Working temperatures of up to 260 °C can be achieved. The HotColumn™ is an optional module for up to six column reactors. Additional accessories include various back pressure regulators (5–50 bar), coil reactors (2–60 mL), etc.

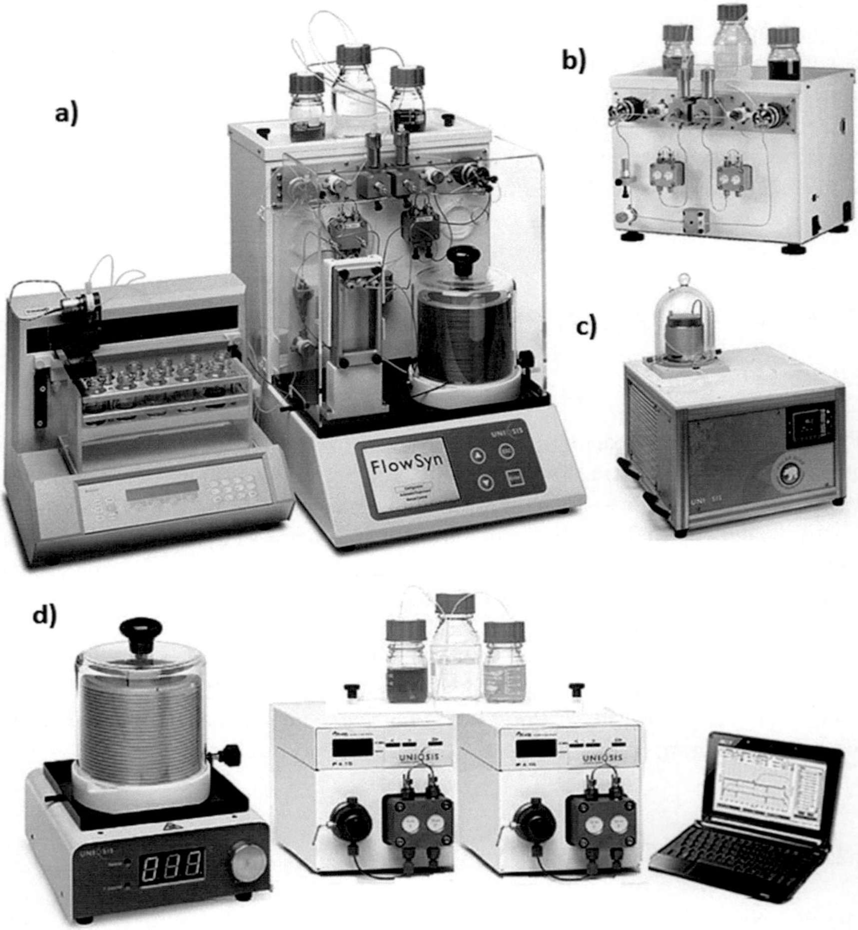

Fig. 2.5 Uniqsis instruments—(**a**) FlowSyn™; (**b**) binary pumping module; (**c**) Polar Bear module; (**d**) FlowStart™

2.2.5 Future Chemistry Holding BV [12]

2.2.5.1 Flow Start Evo

The Flow Start Evo is a compact, stand-alone flow system with various add-on modules (Fig. 2.6). The main module incorporates three syringe pumps (1 µL to 2.9 mL/min) and a microreactor (chip, internal volume ca. 100 µL) holder/heater (up to 140 °C). A photochemistry module as add-on allows irradiation at 250, 295, 365, and 470 nm. The high-temperature module allows working at temperatures between −10 and 200 °C. An additional back pressure regulator can keep an inside

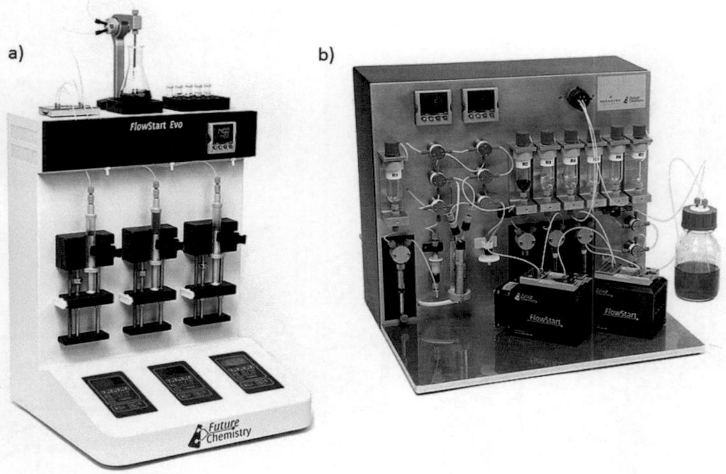

Fig. 2.6 Future chemistry instruments—(**a**) FlowStart Evo and (**b**) FlowStart Expert

system pressure of up to 5 bar. The optional combination with a gas module makes gas/liquid reactions easy and precisely handled. The application of many standard noncorrosive gases is possible. Finally, a computer control of the overall system is also possible.

2.2.5.2 Flow Start Expert

The Flow Start Expert (Fig. 2.6) is an advanced flow setup with fully automated liquid handling and integrated vacuum pump. The integrated automated valves, reagent vials, and sample collectors allow library synthesis. Inert conditions can be realized for sensitive chemistries. The system can be employed for radiopharmaceutical synthesis.

2.2.6 Chemtrix BV [13]

2.2.6.1 Labtrix® Start

The Labtrix® Start is another compact, plug-and-play platform for laboratory flow synthesis using microreactors (Fig. 2.7). The system compromises a combination of two syringe pumps (extendable up to five), microreactor holder/heater, and a temperature controller. The process window of the system ranges form −20 to 195 °C and 0–25 bar pressure. Various chip mixers/microreactors are available. Three different versions can be chosen—standard, flex, and ultraflex. The standard version allows working with basic conditions: the flex, with slightly acidic

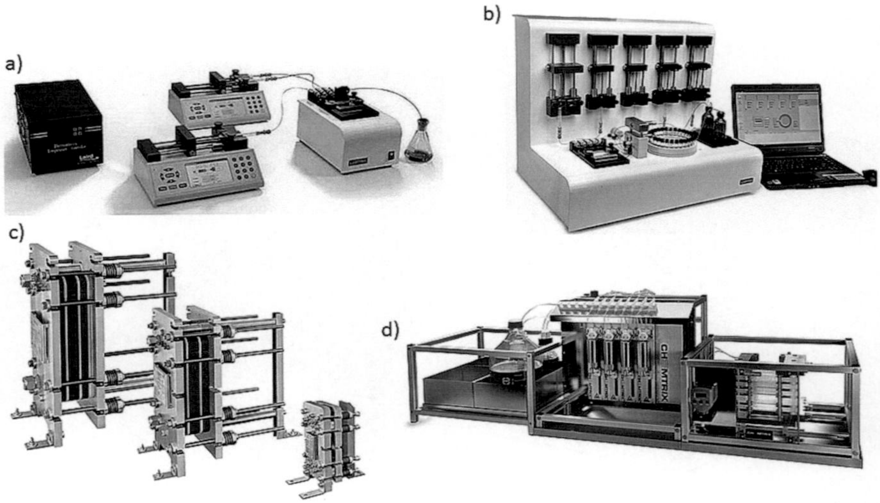

Fig. 2.7 Chemtrix flow instruments—(**a**) Labtrix Start; (**b**) Labrix S1; (**c**) Plantrix reactor modules; (**d**) KiloFlow system

conditions, and the ultraflex, with 70 % nitric acid or 98 % sulfuric acid in −20 to 75 °C range. Upgrades include a flow calculation tool, a catalyst reactor set, a pressure meter set, and an additional feed line.

2.2.6.2 Labtrix® S1

The Labtrix® S1 is a fully automated, plug-and-play platform for laboratory flow synthesis. It has five syringe pumps (1–2.5 mL), two of which can be connected for continuous delivery (Fig. 2.7). Automated sample collection holds up to 30 vials that can be addressed by a selection valve. The temperature/pressure ranges are the same as for the Labtrix® Start system. Three different versions can be obtained here as well—standard, flex, and ultraflex.

2.2.6.3 Kiloflow® and Plantrix®

The Kiloflow® and Plantrix® are glass and ceramic/silicon carbide modular reactors intended for scale-up flow synthesis on kilogram/t scale (Fig. 2.7).

Fig. 2.8 Advion NanoTek system—different modules

2.2.7 *Advion Inc. [14]*

2.2.7.1 NanoTek®

The NanoTek® is a modular microfluidic system developed for radiochemical synthesis of PET and SPECT imaging probes (Fig. 2.8). The system can handle pressure of up to 28 bar and works in the temperature range of −40 to 220 °C. It consists of syringe pumps, a reactor module, and a concentrator/evaporator unit. The system can be extended to allow automated HPLC purifications of the obtained products. Various reactor volumes are available.

2.2.8 *YMC Co. Ltd [15]*

2.2.8.1 The KeyChem Reactors

The KeyChem concept includes two modular microreactor systems for laboratory flow applications—the KeyChem Basic and the KeyChem-L. Both systems are based on the use of syringe pumps in combination with reactor modules (Peltier thermostated) and either manual or computer control. Additionally, the KeyChem Lumino is available. It comprises a micromixer, a thermostat, and a UV LED light source for continuous flow photochemistry experimentation.

2.2.8.2 CYTOS-200 and CYTOS-2000

The CYTOS reactors are aimed at scaling-up synthesis and use of either syringe or piston pumps.

2.2.9 *AM Technology [16]*

2.2.9.1 Coflore® ACR and ATR

The Coflore® systems contain multistage flow reactors that are intended to overcome the problems of slurry and suspension processing under flow conditions while, at the same time, assuring efficient mixing. The patented mixing technique is based on freely moving agitators within each reactor section promoting efficient mixing by lateral shaking of the reactor body. This special action prevents phase separation. The Coflore® ACR system has a small footprint (Fig. 2.9) and fits on a standard laboratory bench [17]. It consists of two parts—the agitator and the exchangeable reactor block. Based on the reactor block, three different versions are available—ACR-20 (10–17 mL reactor volume), ACR-100 (30–90 mL), and ACR-X (countercurrent flow; for reactions or extractions). All of the configurations are designed to withstand temperatures of – 40 °C up to 140 °C and ambient up to 10 bar pressure. The reactor block is available in stainless steel, Hastelloy®, or Teflon® versions.

The Coflore® ATR is an industrial flow reactor system with a capacity range of 0.1–10 L and is based on a tubular design. It can safely manage pressures up to 100 bar and temperatures from −90 °C up to 300 °C (stainless steel or Hastelloy®).

Fig. 2.9 Coflore ACR system from AM Technology

2.2.10 Ehrfeld [18]

2.2.10.1 MMRS®

The Modular MikroReaktionsSystem (MMRS) is highly flexible concept providing more than 60 single modules for assembling a flow process on a laboratory scale [19]. The different modules are mounted together on a metal plate with a variable size. Multistep syntheses are also realizable. Using different sensors and actuators, real-time data useful for the process optimization can be collected easily. The specifications of the system include the following:

- Temperature regimes: −25 °C to 200 °C (−160 °C to 600 °C)
- Working pressures: up to 100 bar
- Flow rates: 0.16–500 mL/min
- Wetted materials: Teflon®, FFKM, stainless steel, and Hastelloy®
- Modules for mixing, emulsifying, heterogeneous/homogeneous synthesizing, photochemistry

  The system requires external pumps.

2.2.10.2 FlowPlate®, ART®, and Miprowa®

The reactors are intended for scale-up purposes on industrial level.

2.2.11 Corning [20]

The Corning flow reactors have a modular chip-based design and are aimed predominantly at large-scale and industrial-scale synthesis. In addition, the low-flow and the G1 systems offer possibilities for smaller-scale syntheses. A version of the G1 reactor was developed to allow photochemical synthesis using UV LED irradiation at 365 and 405 nm. The G3 and G4 are large-scale devices.

2.2.12 Accendo Corporation [21]

2.2.12.1 Conjure™ Flow Chemistry

The Conjure is a fully automated system for continuous synthesis. For library synthesis or screenings, up to 40 different materials can be preloaded. Automated segment preparation allows a broad spectrum of stoichiometries to be tested. The Conjure can be coupled easily with an LC/MS—automated sample preparation, dilution, and injections are possible. Multistep synthesis is also possible. Temperature regimes cover the range of −20 °C to 100 °C [22].

2.2.12.2 Propel™ Flow Chemistry

The Propel is a modular system for flow synthesis specifically designed as a shared resource. The system can hold up to three reactant materials for screening, optimization, and the scale-up of a reaction. The programming interface allows an initial setup of up to nine experiments for optimizing stoichiometry, residence time, and temperature. The Propel system can routinely perform experiments with as little as 20 μL of precious reactants. A scale up to over 100 g is easily achievable. An on-line LC–MS module can be used as an extension.

References

1. H.M. Brennecke, K.A. Kolbe, Ind. Eng. Chem. **48**, 1298 (1956)
2. K. Köll, J. Metzger, Angew. Chem. **90**, 802 (1978)
3. http://www.thalesnano.com
4. I. Kovacs, R.V. Jones, K. Niesz, C. Csajagi, B. Borcsek, F. Darvas, L. Urge, J. Lab. Autom. **12**, 284 (2007)
5. T. Razzaq, T.N. Glasnov, C.O. Kappe, Chem. Eng. Technol. **32**, 1702 (2009)
6. T. Razzaq, T.N. Glasnov, C.O. Kappe, Eur. J. Org. Chem. **9**, 1321 (2009)
7. M. Damm, T.N. Glasnov, C.O. Kappe, Org. Process. Res. Dev. **14**, 215 (2010)
8. http://syrris.com
9. R. Gray, The Biochemist 27, 15 (2005)
10. https://www.vapourtec.com
11. http://www.uniqsis.com
12. http://futurechemistry.com
13. http://www.chemtrix.com
14. http://www.advion.com/products/nanotek/nanotek-liquid-flow-reactor/
15. http://ymc.de/keychem-basic.html
16. http://www.amtechuk.com/Products.aspx
17. D.L. Browne, B.J. Deadman, R. Ashe, I.R. Baxendale, S.V. Ley, Org. Process. Res. Dev. **15**, 693 (2011)
18. http://www.ehrfeld.com/home.html
19. D.A. Snyder, C. Noti, P. Seeberger, Helv. Chim. Acta. **88**, 1 (2005)
20. https://www.corning.com/in/en/innovation/corning-emerging-innovations/advanced-flow-reactors.html
21. http://accendocorporation.com/index_files/Products1.html
22. J.E. Hochlowski, P.A. Searle, N.P. Tu, J.Y. Pan, S.G. Spanton, S.W. Djuric, J. Flow Chem. **1**, 56 (2011)

Chapter 3
Organic Synthesis in Dedicated Continuous Flow Systems

Transition Metal-Catalyzed Carbon–Carbon and Carbon–Heteroatom Bond Forming Reactions

3.1 Suzuki Reaction

The Suzuki reaction (the coupling of an aryl halide with a boronic acid in the presence of a palladium catalyst) is one of the most widely used cross-coupling reactions in modern organic synthesis—in the research laboratory as well as on industrial scale. Typically, high-speed Suzuki reactions are performed at elevated temperatures. Only a few continuous flow examples have been demonstrated using either homogenous or heterogeneous Pd catalysts.

A team of chemists at ThalesNano Nanotechnology used an X-Cube flow system to perform various reactions including a Suzuki coupling back in 2007 to demonstrate the wide applicability of the instrument [1]. Later on, the X-Flash was employed in a two-step continuous flow process for the preparation of 2-amino-4′-chlorobiphenyl—an important intermediate in the synthesis of the fungicide Boscalid®—being produced by BASF on more than a 1000 tonnes/year scale [2]. A high-temperature Suzuki–Miyaura cross-coupling reaction, using tetrakis (triphenylphosphine)palladium under homogeneous flow conditions, delivers the central biaryl unit of the Boscalid®.

Accordingly, 1-chloro-2-nitrobenzene is coupled with 4-chlorophenylboronic acid in the microtubular flow reactor at 160 °C using a *tert*-butanol/water/potassium *tert*-butoxide solvent/base system to obtain the corresponding biphenyl in high yield. The Pd catalyst had to be removed via a QuadraPure resign prior to the following highly chemoselective heterogeneous Pt/C-catalyzed nitro group reduction (a coupled flow hydrogenation process) and an amide bond formation steps to afford the desired active molecule (Scheme 3.1). The Suzuki coupling was also used as a test- reaction by Leadbeater et al. to illustrate the simplicity of transferring of a microwave batch process to a flow procedure [3].

Alcazar et al. were using a Vapourtec® flow system and reported a continuous flow Suzuki protocol as well [4]. Here, a solid-supported Pd catalyst was employed. Although it is commonly accepted that supported metal catalysts are not optimal to

T. Glasnov, *Continuous-Flow Chemistry in the Research Laboratory*,
DOI 10.1007/978-3-319-32196-7_3

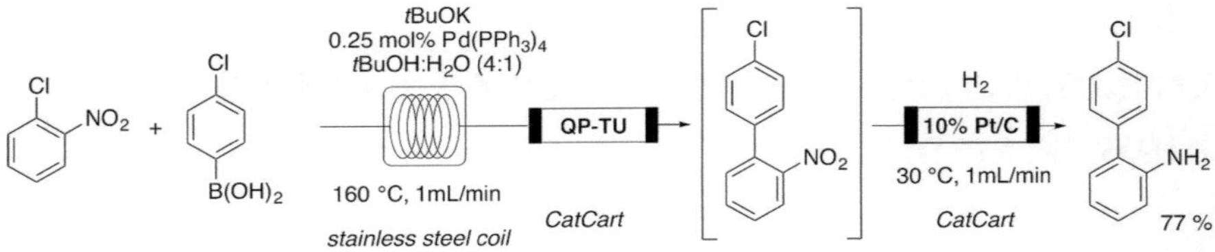

Scheme 3.1 Flow synthesis of 2-amino-4′-chlorobiphenyl

use under flow conditions [5, 6], it was demonstrated that the used SiliaCat DPP–Pd is robust enough to allow a successful 8 h long continuous run without significant decrease in conversion. A small library of differently substituted biaryls was thus easily generated.

With similar purpose and instrumental setup, Frost and coworkers reported on the preparation and application of a polymer-encapsulated Pd(0) catalyst. The robustness and reusability of the latter was thoroughly examined. The prepared catalyst could be used for over 50 h with no appreciable decrease in activity [7].

A homogenous C–C coupling process was evaluated for the preparation of the biaryl unit of Odanacatib®—an orally administered very active and selective cathepsin K inhibitor in development for the treatment of postmenopausal osteoporosis [8]. Initially, a microwave batch optimization showed the way to the optimal conditions for the coupling reaction, which were later on transformed into a continuous flow process. The starting enantiomerically pure alcohol was obtained in an asymmetric enzymatic batch reduction process prior the Suzuki–Miyaura coupling. Using only 0.2 mol% of homogenous $Pd(PPh_3)_4$, 72 % overall yield from the corresponding biaryl alcohol at 110 °C and 5 min residence time in the stainless steel flow reactor were obtained (Scheme 3.2).

The X-Cube's CatCart concept was employed by Gordon and coworkers to prepare furan-containing biaryls [9]. Here, immobilized palladium was used as the optimal heterogeneous catalyst source for the coupling process. A number of polymer-supported catalyst were tested—FC1032™, FC1001™, FC1007™, and $PdCl_2(PPh_3)_2$. Various tetrabutylammonium salts were also screened as reaction additives. Cross-couplings of 5-formyl-2-furanylboronic acid with numerous aryl bromides were performed under continuous flow conditions using the $(Bu)_4NF$/FC1032™ combination. Deactivated aryl bromides and activated aryl chlorides required the $(Bu)_4NOAc/PdCl_2(PPh_3)_2$ additive/catalyst combination. The optimized conditions were tested with iodides, bromides, and chlorides demonstrating the general applicability of the flow protocol. Using methanol as solvent at 120 °C and in only short residence time inside the CatCart, the reaction required recycling to obtain full conversion in many cases. Nevertheless, in most of the cases, nearly quantitative conversions could be obtained. Evaluation of the Pd leaching via ICP-MS showed leaching of the catalyst, as one could expect, although on a rather negligible level.

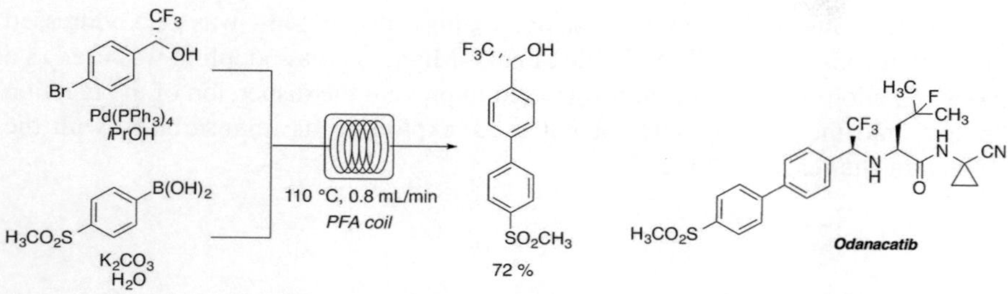

Scheme 3.2 Biaryl unit for the synthesis of odanacatib

The efficiency, durability, and metal leaching of immobilized di- and triarylphosphine Pd catalysts under continuous flow conditions have been recently investigated in detail using the CatCart concept of the X-Cube flow reactor [5]. These key parameters determine the choice of a catalyst for performing metal-catalyzed cross-coupling reactions under flow conditions. This comparative investigation included some of the most common immobilized phosphine-based Pd catalysts—$Pd(PPh_3)_4$ (polymer bound), FC1001™, EnCat™ TPP30, and SiliaCat DPP–Pd. The efficiency, recyclability, and leaching resistance of each of the catalysts have been investigated in detail using a set of literature-based conditions protocols. Based on the obtained results using various catalysts in a sample Suzuki–Miyaura reaction, it could be demonstrated that in many cases, a more appropriate approach would be the use of homogenous catalytic systems to ensure reproducibility, as the reaction conditions are easily and accurately adjustable to the process requirements. Interestingly, the SiliaCat DPP–Pd showed increased resistance to leaching. Further investigations in this direction might lead to much more robust supported catalysts in the future.

The probably simplest supported Pd catalyst—Pd on charcoal (Pd/C)—was also evaluated as a solid-supported source for Pd in a continuous flow Suzuki–Miyaura coupling [10]. An H-Cube flow unit, originally designed for performing flow hydrogenations, was used in this study without utilizing the hydrogen generation cell. CatCarts (30 mm in length, ca. 0.3 mL volume) loaded with Pd/C were employed. Various iodo- and bromoaryls were tested using ethanol–water (1:1) as a solvent mixture and Na_2CO_3 as a base, affording yields above 78 % in a single pass through the cartridge with the catalyst at 1 mL/min flow rate (several seconds residence time). Most interestingly, no leaching of Pd species could be detected (<1 ppm detection limit, atom absorption spectroscopy). The authors speculated that the low reaction temperature (25 °C), the flow rate, and the low substrate concentration (0.05 M) were responsible for this observation.

In yet another study on immobilized Pd catalysts, new silica-supported Pd–NHC complexes were evaluated under flow conditions using the Vapourtec® flow system [11]. The catalyst was prepared and examined in the Suzuki–Miyaura coupling of a range of bromo- and chloroaryls with obtained conversions between 55 and 92 %. Flow experimentation showed moderate conversions for 2 h of uninterrupted processing.

Another issue in continuous flow processing—dispersion—was also addressed in a very recent report [12]. A simple Suzuki–Miyaura cross-coupling was used as a model reaction. Perfluorodecalin was used to prevent the dispersion of the reaction "plugs" into the bulk organic solvent used, exploiting its immiscibility with the solvent of the reaction "plugs."

3.2 Heck Reaction

The Heck reaction, a palladium-catalyzed vinylic substitution, is an example of a Pd-catalyzed cross-coupling of alkenes and organohalides/pseudohalides in the presence of a base. The first report of a Heck reaction in a commercially available system emerged in the year 2005 using an Ehrfeld modular system [13]. Ethyl acrylate was reacted with phenyl iodide using Et_3N as an organic base. 10 % Pd/C as a solid-supported Pd catalyst was employed. The 30 min residence time at 130 °C assured a 95 % yield of ethyl cinnamate.

Several years later, an in-depth evaluation of a very similar reaction under microwave batch and continuous flow conditions using a CEM Discover and X-Flash reactors was disclosed by Kappe et al. [5]. The effects of temperature, time, types of catalysts (heterogeneous versus homogenous), and additives on the reaction were examined as well as catalyst leaching. A follow-up of this study, focused on the use of various heterogeneous Pd catalyst and their performance under flow conditions, was recently disclosed by the same group.

The synthesis of nabumetone (NSAID) and related 4-aryl-butanones was assessed under continuous flow conditions [14]. The synthesis comprises two consecutive steps—a coupling and a double-bond hydrogenation. For the first step, three different reactions were evaluated—Heck cross-coupling, Wittig olefination, and an aldol-type condensation. Initially, the reactions were evaluated under microwave batch conditions and the best one transformed into a continuous flow process. The Heck reaction was realized in an X-Flash system working at preset conditions of 180 °C and 10 min residence time. Full substrate conversion was observed and 67 % yield isolated (Scheme 3.3).

Although successful, due to the rather low yield and selectivity achieved, in combination with the required purification of the crude reaction mixture and the high costs of catalyst and starting materials (aryl iodides), the process was not advanced further.

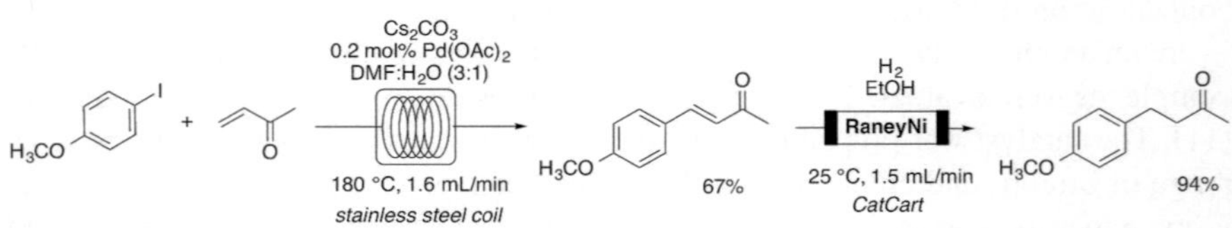

Scheme 3.3 Mizoroki–Heck and hydrogenation reactions under flow conditions

A decarboxylative Heck reaction of 2,6-dimethoxybenzoic acid and methyl acrylate in the presence of oxygen gas was performed in a Vapourtec system using the recently developed "tube-in-tube" reactor concept, which allows the use of reactive gases under flow conditions [15]. Preliminary results of the high-temperature process were obtained using a microwave instrument. Mimicking the optimal batch conditions, the process could be reproduced in the selected continuous flow setup, delivering comparable results (86 % versus 90 % in batch).

Optimized Heck coupling conditions, relying on minimal amounts (0.05 mol%) of $Pd(OAc)_2$ as a catalyst, were established lately by Price and coworkers [16]. The optimization work and the scaled reaction were realized on a Conjure flow system and supported by design of experiment (DoE) studies. Thus, Pd amounts of 500 ppm were identified as optimal whenever using iodides as starting materials. A reaction time of 5 min at 180–200 °C proved to be optimal.

An X-Cube reactor was also implemented in a study of supported ionic liquid phase Pd catalysts [17]. The Heck coupling of methyl acrylate and iodobenzene was used as a model reaction for the chemical properties of the newly prepared catalysts. The used conditions for the heterogenization of the catalyst were identified as the crucial factor for the effectiveness of the Heck reaction. Several operation hours of selective Heck coupling could be also achieved under continuous flow conditions, whereby a strong base in ethanol as the reaction solvent was used.

3.3 Sonogashira Reaction

The Sonogashira reaction, a palladium/copper-catalyzed coupling of terminal acetylenes with aryl and vinyl halides, is well known as a reliable method for the synthesis of unsymmetrical alkynes.

In 2011 a commercially available copper tube reactor was described to catalyze various reactions including a Sonogashira coupling without the need of additional metals, ligands, or further reagents in very high yields [18]. A general protocol was developed, using dimethylformamide as the solvent of choice and tetrabutylammonium acetate as the base. In the case of less reactive substrates—bromobenzenes or trimethylsilylacetylene—catalytic amounts of Pd were required. Interestingly, homocoupling of the alkyne was not observed (Glaser–Hay reaction). Isolated yields as high as 94 % were easily achievable.

An H-Cube flow hydrogenation reactor with fixed-bed catalyst was used in "non-hydrogenation" mode for the performance screening of various immobilized palladium catalyst in the Sonogashira coupling [19]. $PdCl_2(PPh_3)_2$, FC1001, FC1007, and Pd/C were successfully tested. However, the authors did not investigate the leaching properties and the performance—crucial characteristics for a flow process based on immobilized catalysts.

A continuous flow approach was applied for the synthesis of various fluorinated alkynyl arenes and heteroarenes including a homologue of the ^{18}F labeled imaging

Scheme 3.4 Sonogashira coupling with Pd/Cu-mixed catalyst and basic conditions

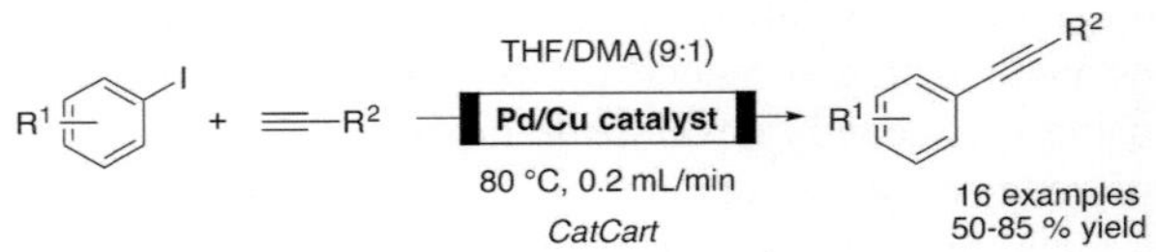

agent Fallypride® [20]. A Labtrix flow instrument was used to obtain the best reaction conditions for a copper-free Sonogashira-type coupling. Readily available building blocks were successfully derivatized with fluoroalkyl side chains in short reaction times (<10 min).

As part of an industrial discovery project of novel Abl kinase inhibitors, a Sonogashira coupling was applied to generate diversity using 27 different aromatic alkynes [21]. The copper tube reactor concept was successfully applied to provide a library of Sonogashira products in 20 min reaction time and 150 °C temperature. For this synthetic purpose, a Vapourtec flow system was chosen.

An intriguing idea to explore Sonogashira reactions under nonbasic conditions for base-sensitive substrates was recently disclosed [22]. A combination of palladium and copper catalyst was needed to realize the intended studies. Along the lines of the idea, a reaction mixture containing substituted iodobenzene and aryl acetylene in a dry solvent mixture of THF-DMA (9:1) was passed over a CatCart filled with Escat1241™—a mixture of 5 % Pd/Al_2O_3 and 0.1 % Cu_2O/Al_2O_3 in 17:1 ratio (Scheme 3.4). However, spatial separation of the two catalysts led to immediate disruption of the catalytic sequence, and no coupling products were observed.

3.4 Negishi Reaction

The Negishi cross-coupling is a another powerful C–C bond forming reaction whose popularity remained lower as compared to other cross-coupling methods in part due to the involvement of the required but less available organozinc species. These are also problematic in terms of reproducibility and general sensitivity.

To overcome these obstacles, a continuous flow process was designed, using an activated packed bed of metallic zinc to prepare the corresponding organozinc reagents in situ, followed by the immediate subsequent use in a Negishi coupling [23]. A single column of packed zinc provided excellent yields of organozinc halides that were immediately involved downstream in subsequent Negishi cross-coupling process in a second DPP–Pd-packed column. Importantly, several factors had to be considered when working with a Zn-packed column—particle size tuning, activation of the metal, column packing, and temperature. With a single packed column, containing 12 g of zinc, 150 mL of a 0.5 M solution of an organozinc halide could be prepared and used in a single run (Scheme 3.5).

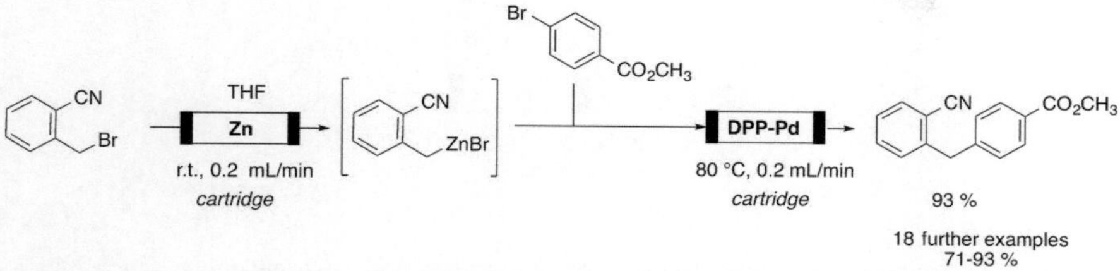

Scheme 3.5 Organozinc reagents generation and application in a Negishi coupling

3.5 Carbonylations

Csajagl et al. took advantage of the safe handling of reactive gases in contained flow environment and reported on a valuable aminocarbonylation reaction using a supported Pd catalyst (Scheme 3.6) [24]. The most remarkable feature of this protocol is the safe handling of extremely toxic carbon monoxide gas, which is required in carbonylation chemistry.

The experimental procedure was used for the generation of dicarboxylic acid monoamides possessing valuable pharmacological properties. The final conditions used a combination of supported Pd catalyst (0.4 g in a CatCart) in the presence of triethylamine (2 eq) and tetrahydrofuran as a solvent. Safely introducing the CO gas in closed environment at 100 °C and 30 bar pressure, aminocarbonylation was achieved in only 2 min time. Various iodo- and bromocarboxylic acids were reacted under optimized reaction conditions to deliver good yields of the desired monoamides. Interestingly, when working with aryl iodides and bromides under similar conditions using homogeneous $Pd(PPh_3)_4$ catalyst, mixtures of mono- and dicarbonylation products were observed [25].

An alkoxycarbonylation with carbon monoxide and aryl iodides to generate the corresponding aromatic acid esters was achieved on a Vapourtec system by Mercandante and Leadbeater [26]. Two sequential "tube-in-tube" reactors were required to obtain optimal results—91–99 % conversion (NMR) for the eight described examples. A catalytic system with 0.5 mol% $Pd(OAc)_2$ and 1.1 eq of DBU as a base in the corresponding alcohol as a solvent were able to effectively convert the starting material into product at 120 °C reaction temperature. The process was evaluated also on Uniqsis FlowSyn reactor [27]. The subsequent simplification of the continuous flow setup did not require a "tube-in-tube" reactor setup but a simple T-shaped mixer to introduce the CO gas into the reaction mixture. Recently the group of Steven Ley reevaluated these reaction types—alkoxy-, hydroxy-, and aminocarbonylations [28]. Further substrates were evaluated, including aliphatic, aromatic bromides and iodides as well as intramolecular versions of the reaction. The "tube-in-tube" concept was applied here as well.

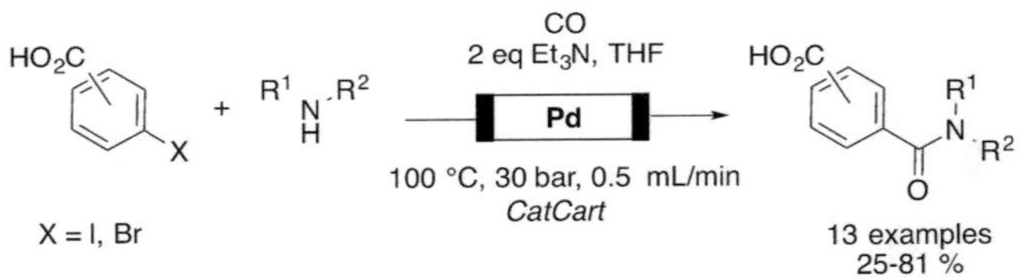

Scheme 3.6 Aminocarbonylation reaction with CO gas in a flow environment

3.6 Carbon–Heteroatom Coupling Reactions

In 2009 Eycken et al. elaborated a continuous flow procedure for the copper(II)-mediated *N*- and *O*-arylations of various compounds with arylboronic acids [29].

The Chen–Lam–(Evans) cross-coupling leads to an efficient C(aryl)–O and C (aryl)–N bond formation using arylboronic acids in the presence of a base and copper acetate as a catalyst under much milder conditions than the Buchwald–Hartwig coupling method. Initial optimization of possible reaction conditions in a CYTOS College system provided best results for the desired C–N coupling when using dichloromethane as a reaction solvent at room temperature, triethylamine/pyridine mixture as a base, and copper acetate as the catalyst. Several secondary aromatic amines could be obtained on a gram scale in 56–73 % isolated yields. The scope was broadened to aliphatic amines and amides with *N*-phenylated caprolactam and cyclohexylamine without the need of reoptimization. To obtain diaryl ethers from the corresponding phenols, the reaction conditions had to be changed to employ dimethylformamide as solvent and 130 °C as reaction temperature. These changes made it possible to achieve full conversions on a gram scale for the tested substrates (Scheme 3.7). A year later, the same authors reported an A^3-coupling reaction for the synthesis of dibenzazocines and dibenzazepines. The coupling of an aldehyde, amine, and a terminal alkyne in the presence of copper catalyst led to the desired functionalized seven-membered heterocycles (Scheme 3.8). A protocol involving both microwave batch synthesis and subsequent continuous flow process was developed. A rapid reaction, requiring only 5 min of residence time, was made possible by passing the reaction mixture over a Cu/C-packed CatCart-cartridge at 150 °C temperature. In a direct comparison, an identical 79 % yield was obtained in both microwave batch and continuous flow experiments. At low flow rates (below 1.5 mL/min), a Glaser–Eglinton–Hay coupling was observed, as could be expected in the presence of the copper catalyst and the terminal alkyne [30].

Another copper(II)-catalyzed C–O coupling process was recently evaluated under flow conditions to generate hitherto unreported unsymmetrical acetal scaffolds (Scheme 3.9) [31]. Using *tert*-butyl hydroperoxide (TBHP) was the key requirement for the catalytic process although the combination of a peroxide with ethers at higher temperatures is potentially hazardous. In the final continuous flow process, two reagent streams—one containing the substrate, catalyst, and the ether, as a solvent and the second with THBP as a commercial decane solution—were

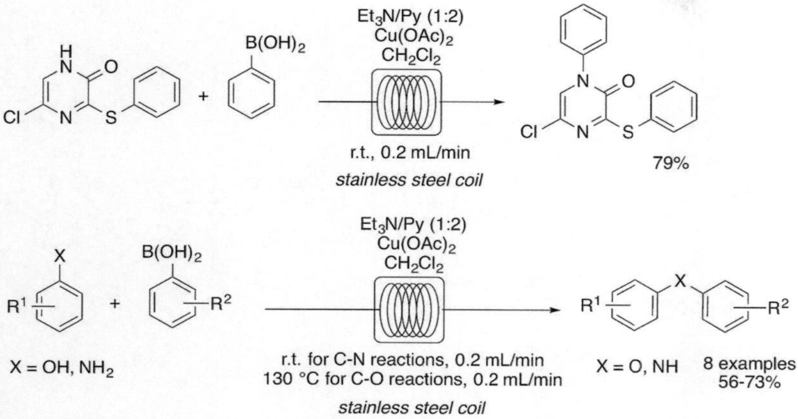

Scheme 3.7 Copper-catalyzed C–N and C–O coupling reactions in flow

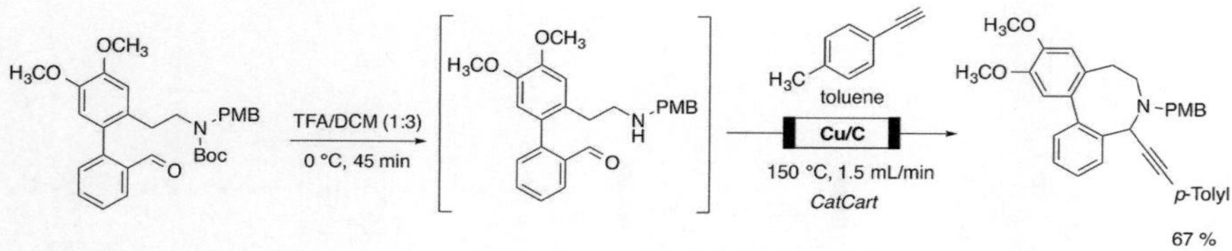

Scheme 3.8 Flow synthesis of 5,6,7,8-tetrahydrodibenzo[c,e]azocines via the A^3-coupling reaction

processed at 130 °C for 20 min to provide a range of products in similar yields as in the microwave batch experiments.

The gold-catalyzed alkylation of various amines with alcohols was disclosed by Hii and coworkers in 2012 [32]. A synthetic strategy relying on the initial catalytic oxidation of the alcohol to aldehyde, condensation with the amine to an imine and reduction of the latter to the final alkylated amine was considered. This can be achieved by a catalytic H_2 transfer from the alcohol to the imine via metal hydride intermediates ("borrowing hydrogen"). The most relevant and effective known catalysts to affect such a process are based on iridium and ruthenium. Several variations of the catalyst also exist. The Hii group exploited a commercial Au/TiO_2 heterogeneous catalyst, which proved to be sufficiently active and selective to catalyze the alkylation. The optimal process required 180–200 °C reaction temperature and 1–7 h recycling of the reaction mixture over the catalyst bed to convert the mixture of the corresponding alcohol and the amine into the alkylated products with high conversions and selectivity. Various mechanistic aspects were evaluated especially the effect of water on the reaction progress at elevated temperatures was observed.

As should be evident for flow processes, the use of solids is highly problematic due to the danger of clogging. In certain cases, the insolubility of some compounds

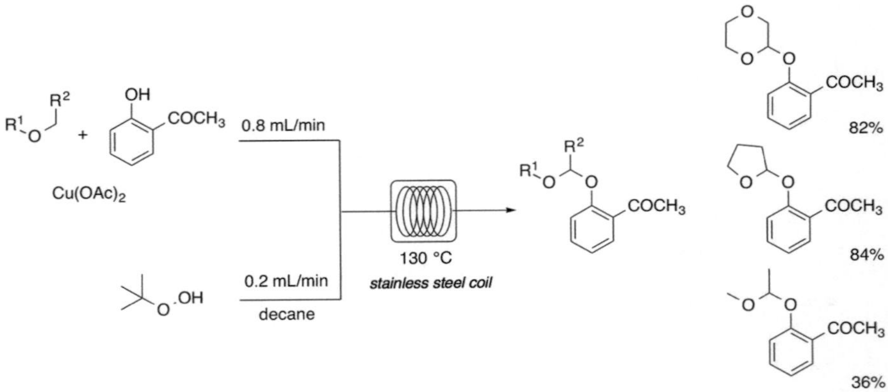

Scheme 3.9 Two-feed continuous flow oxidative C–O coupling of 2-hydroxyacetophenone with different ethers

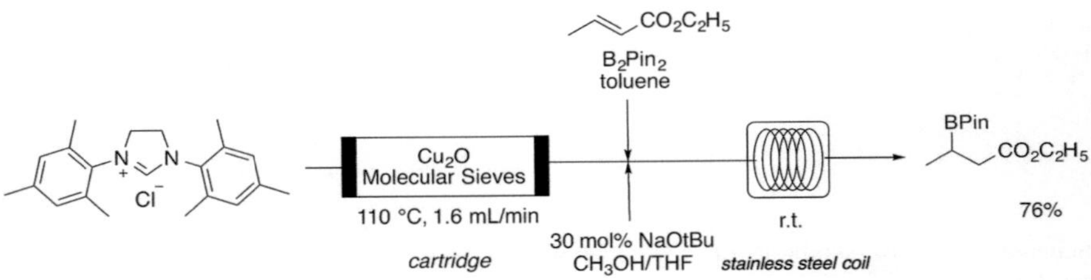

Scheme 3.10 Synthesis and application of NHC–CuCl catalyst in a β-borylation reaction

can be turned into an advantage: copper complexes were produced using a flow chemistry approach whereby soluble ligands were flowed through a packed bed of insoluble metal source and consequently used in downstream reactions, thus representing an effective alternative to glove box and Schlenk working techniques [33]. The synthesis of air-sensitive *N*-heterocyclic carbene (NHC)–copper chloride complexes from insoluble Cu_2O and NHC precursors was accomplished by employing a packed bed of Cu_2O downstream of the pumps to avoid clogging issues. Depending on the dead volume of the used columns, amounts of up to 11 g of the Cu–NHC complexes were easily synthesized at 110 °C within 1 min, whereas batch experiments failed. A subsequent β-borylation (C–B bond formation) was used to demonstrate the abilities of the Cu–NHC complex generating system. The β-borylation step was realized in an adjacent coil thermostated to 0 °C. Eleven grams of corresponding β-borylated product were obtained in a 40 min run (Scheme 3.10).

3.7 Miscellaneous

A continuous flow synthesis of a carbon-based molecular cage macrocycle via a threefold homocoupling reaction was achieved by employing a CuCl–Cu(OAc)$_2$ catalyst mixture in a Glaser–Eglinton–Hay coupling reaction [34]. Shape-persistent cage molecules are of interest in may research areas because of their unique physical properties. Here, the Breslow modification of the Glaser–Eglinton–Hay conditions was found to be the best synthetic option. Although the flow approach is easily scalable, it should be mentioned that a batch procedure provided essentially the same yield (20 % batch versus 21 % in flow). The standard procedure uses two reagent streams: a pre-mixed solution of 21 eq anhydrous Cu(OAc)$_2$ and 32 eq of CuCl in dry pyridine and a 1 mmol solution of a “half cage” in dry pyridine. These are combined in a T-mixer before entering the reaction coil preheated to 70 °C to achieve the threefold coupling (Scheme 3.11).

A Cu–Pd catalyst system was used for the convenient decarboxylative cross-coupling of various aromatic carboxylic acids and aryl triflates to prepare numerous biaryls [35]. In an optimization study, the temperature, the catalyst amount, and the Cu/Pd ratio were examined. With the best conditions at hand—5 mol% Cu catalyst, 2 mol% Pd catalyst, 170 °C reaction temperature, and 1 h residence time—the expected biaryls were obtained in 5–82 % yields.

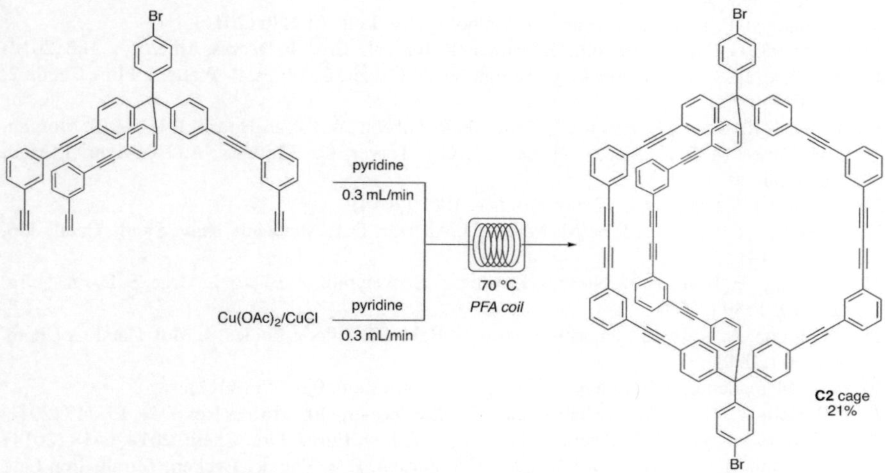

Scheme 3.11 Flow synthesis of a C2 cage

References

1. I. Kovacs, R.V. Jones, K. Niesz, C. Csajagi, B. Borcsek, F. Darvas, L. Urge, J. Lab. Autom. **12**, 284 (2007)
2. T. Glasnov, C.O. Kappe, Adv. Synth. Catal. **352**, 3089 (2010)
3. C.B. Kelly, C.X. Lee, N.E. Leadbeater, Tetrahedron Lett. **52**, 263 (2011)
4. J. de M. Munoz, J. Alcazar, A. de la Hoz, A. Diaz-Ortiz, Adv. Synth. Catal. **354**, 3456 (2012)
5. T.N. Glasnov, S. Findenig, C.O. Kappe, Chem. Eur. J. **15**, 1001 (2009)
6. R. Greco, W. Goessler, D. Cantillo, C.O. Kappe, ACS Catal. **5**, 1303 (2015)
7. W.R. Reynolds, P. Plucinski, C.G. Frost, Catal. Sci. Technol. **4**, 948 (2014)
8. R. de O. Lopes, A.S. de Miranda, B. Reichart, T. Glasnov, C.O. Kappe, R.C. Simon, W. Kroutil, L.S.M. Miranda, I. C.R. Leal, R.O.M.A. de Souza, J. Mol. Catal. B **104**, 101 (2014)
9. T.N. Trinh, L. Hizartzidis, A.J.S. Lin, D.G. Harman, A. McCluskey, C.P. Gordon, Org. Biomol. Chem. **12**, 9562 (2014)
10. T. Hattori, A. Tsubone, Y. Sawama, Y. Monguchi, H. Sajiki, Catalysts **5**, 18 (2015)
11. A. Martinez, J.L. Krinsky, I. Penafiel, S. Castillon, K. Loponov, A. Lapkin, C. Godard, C. Claver, Catal. Sci. Technol. **5**, 310 (2015)
12. N. Hawbaker, E. Wittgrove, B. Christensen, N.W. Sachl, D.G. Blackmond, Org. Process Res. Dev. 10.1021/op500360w (2015)
13. D.A. Snyder, C. Noti, P.H. Seeberger, Helv. Chim. Acta **88**, 1 (2005)
14. M. Viviano, T.N. Glasnov, B. Reichart, G. Tekautz, C.O. Kappe, Org. Process Res. Dev. **15**, 858 (2011)
15. D.M. Rudzinski, N.E. Leadbeater, Green Process. Synth. **2**, 323 (2013)
16. P. Cyr, S.T. Deng, J.M. Hawkins, K.E. Price, Org. Lett. **15**, 4342 (2013)
17. B. Urban, D. Sranko, G. Safran, L. Ürge, F. Darvas, J. Bakos, R. Skoda-Földes, J. Mol. Catal. A Chem. **395**, 364 (2014)
18. Y. Zhang, T.F. Jamison, S. Patel, N. Mainolfi, Org. Lett. **13**, 280 (2011)
19. B. Borcsek, G. Bene, G. Szirbik, G. Dorman, R. Jones, L. Ürge, F. Darvas, Arkivoc, **v**, 186 (2012)
20. M.S. Placzek, J.M. Chmielecki, C. Houghton, A. Calder, C. Wiles, P. Watts, J. Flow Chem. **2**, 46 (2013)
21. B. Desai, K. Dixon, E. Farrant, Q. Feng, K.R. Gibson, W.P. van Hoorn, J. Mills, T. Morgan, D.M. Parry, M.K. Ramjee, C.N. Selway, G.J. Tarver, G. Whitlock, A.G. Wright, J. Med. Chem. **56**, 3033 (2013)
22. S. Voltova, J. Srogl, Org. Chem. Front. **1**, 1067 (2014)
23. N. Alonso, L.Z. Miller, J. de M. Munoz, J. Alcazar, D.T. McQuade, Adv. Synth. Catal. **356**, 3737 (2014)
24. C. Csajagi, B. Borcsek, K. Niesz, I. Kovacs, Z. Szekelyhidi, Z. Bajko, L. Ürge, F. Darvas, Org. Lett. **10**, 1589 (2008)
25. J. Balogha, A. Kuika, L. Ürge, F. Darvas, J. Bakos, R. Skoda-Földes, J. Mol. Catal. A Chem. **302**, 76 (2009)
26. M.A. Mercadante, N.E. Leadbeater, Org. Biomol. Chem. **9**, 6575 (2011)
27. C.B. Kelly, C.X. Lee, M.A. Mercadante, N.E. Leadbeater, Org. Process Res. Dev. **15**, 717 (2011)
28. U. Gross, P. Koos, M. O'Brien, A. Polyzos, S.V. Ley, Eur. J. Org. Chem. **2014**, 6418 (2014)
29. B.K. Singh, C.V. Stevens, D.R.J. Acke, V.S. Parmar, E.V. Van der Eycken, Tetrahedron Lett. **50**, 15 (2009)
30. J.B. Bariwal, D.S. Ermolatev, T.N. Glasnov, K. Van Hecke, V.P. Mehta, L. Van Meervelt, C.O. Kappe, E.V. Van der Eycken, Org. Lett. **12**, 2774 (2010)
31. G.S. Kumar, B. Pieber, K.R. Reddy, C.O. Kappe, Chem. Eur. J. **18**, 6124 (2012)
32. N. Zotova, F.J. Roberts, G.H. Kelsall, A.S. Jessiman, K. Hellgardt, K.K.M. Hii, Green Chem. **14**, 226 (2012)
33. S.M. Opalka, J.K. Park, A.R. Longstreet, D.T. McQuade, Org. Lett. **15**, 996 (2013)
34. M. Kitchin, K. Konstas, C.J. Sumby, M.L. Czyz, P. Valente, M.R. Hill, A. Polyzos, C.J. Doonan, Chem. Commun. **51**, 14231 (2015)
35. P.P. Lange, L.J. Gooßen, P. Podmore, T. Underwood, N. Sciammetta, Chem. Commun. **47**, 3628 (2011)

Chapter 4
Organic Synthesis in Dedicated Continuous Flow Systems

Rearrangement Reactions

4.1 Curtius Rearrangement

The Curtius rearrangement is synthetically a very useful transformation for thermally converting acyl azides into isocyanates. As a direct chemical transformation, the process is highly valuable and includes the facile generation of the amide bonds in many pharmaceutical compounds and natural products. As it relies on the use of azides and acyl azides, it must be considered a very hazardous process, and thus process safety is of utmost importance. Accordingly, Ley and coworkers reported on the clean and reliable application of the Curtius rearrangement in the contained environment of a flow process using a commercial system [1]. A substrate stream containing an acid, triethylamine, and a nucleophile was combined with another stream containing diphenylphosphoryl azide in acetonitrile and pumped into a heated zone, thermostated to 120 °C. After a residence time of 20–50 min, the corresponding products were collected and isolated. The in situ-obtained isocyanates were directly transformed into a small selection of carbamates, ureas, semicarbazides, and related compounds in good yields (75–95 %) (Scheme 4.1). In-line scavenging was applied as a method of product purification.

Shortly after, the same group reported a related flow process, albeit implementing a heterogeneous azide source to generate corresponding acyl azides [2]. For this purpose, macroporous Merrifield-type monoliths were prepared inside a 15 × 100 mm Omnifit glass column and derivatized with azide functionalities to afford the azide monoliths with 2 mmol/g loading of azide. Monoliths with 3, 7 and 15 mmol/g azide loading per cartridge were then used for the synthetic procedures. For the generation of acyl azides, a stream of the corresponding acyl chlorides in acetonitrile was passed over the azide monoliths at room temperature. The generated acyl azides were then either isolated or easily transformed into the corresponding carbamates or ureas utilizing the procedure described above. Based on the collected experience in the continuous flow Curtius rearrangement and on the search for synthetic routes to new modulators of the histone reader

T. Glasnov, *Continuous-Flow Chemistry in the Research Laboratory*,
DOI 10.1007/978-3-319-32196-7_4

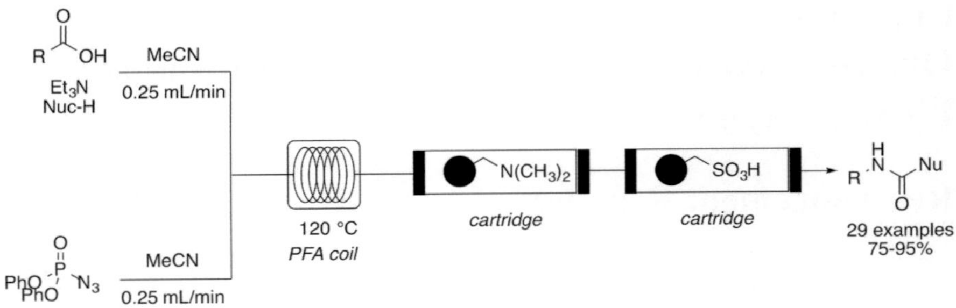

Scheme 4.1 Continuous flow Curtius rearrangement and in-line purification with scavenger resigns

BRD9, the Ley group developed recently the synthesis of a substituted dichloropyrazine on 40 mmol scale which is the key building block for the generation of various histone reader BRD9 modulators [3].

4.2 Rearrangement of Cyclobutanones

Thermal rearrangement of cyclobutanones via a vinylketene intermediate is an established and useful method for the preparation of various densely substituted polyaromatic and heteroaromatic ring systems.

The continuous flow synthetic approach delivered impressive results in the synthesis of the marine natural product cribrostatin, possessing antimicrobial and anticancer activity [4]. A number of related benzoquinones was conveniently prepared in yields above 84 % at 150 °C in dioxane as a solvent. Having the best reaction conditions at hand, the synthesis of cribrostatin was achieved in four steps, with the last comprising a thermal rearrangement of the corresponding heteroaryl cyclobutanone. Processing the dioxane solution at 110 °C for an hour with subsequent exposure to air resulted in 90 % isolated yield of the natural product (Scheme 4.2). Additionally, DFT and ab initio calculations in combination with the Hammett relationship for the reaction improved the understanding of the process. Later, the same group reported also on the synthesis of another natural product—(−) mansonone B—that could be prepared in similar fashion over four consecutive steps [5]. The corresponding cyclobutanone was converted into hydroquinone in 85 % yield, in which an additional deprotection step provided the final product. In addition, the effects of organoytterbium additions on the preparation of other useful cyclobutanones were also evaluated, and an interesting example of a photochemical rearrangement in flow to obtain a furanone in 81 % was demonstrated.

A related rearrangement with ethynylcyclobutanone was evaluated under similar conditions [6]. However, comparative studies with conventional and microwave heating revealed that the microwave batch conditions appear to be the optimal choice in this case. Rationalization for the better yields in the microwave experiments was not provided.

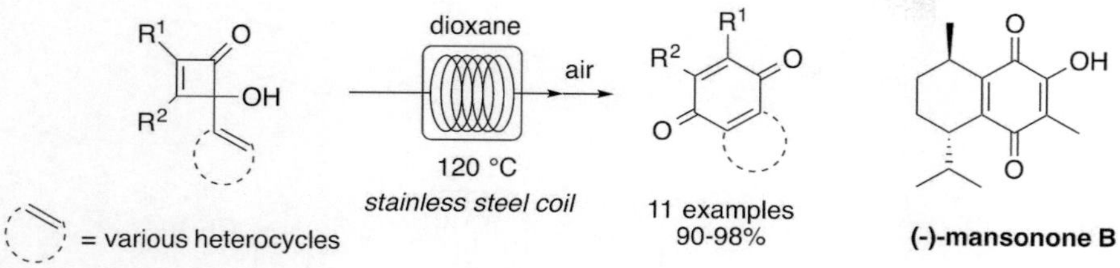

Scheme 4.2 Thermal rearrangement of cyclobutanones

4.3 Miscellaneous

The Hofmann rearrangement is a unique way to convert primary amides into the corresponding carbamates via C–C to C–N bond rearrangement. Utilizing the NanoTek flow platform, Ley and coworkers established a simple flow procedure to solve potential toxicity issues with the use of bromine and bromine reagents at the high reaction temperatures required in these reactions (Scheme 4.3) [7].

Commercially available aromatic amides were used as starting materials, and depending on the substitution, pattern yields in the range of 32–80 % were achieved. Using a FlowSyn instrument, the obtained results could be reproduced and the reaction was scaled up to 1 g.

A modified Beckmann rearrangement—the transformation of oximes into formamides—has been reported to provide primary amides when performed in the presence of copper catalysts [8]. Conventional and microwave batch conditions with $Cu(OAc)_2$ in toluene as solvent afforded a variety of primary amides. Based on the obtained results, a heterogeneous catalyst—CuO/ZnO—was also successfully tested under batch and flow conditions. Working at 150 °C on an X-Cube system, an 88 % yield of the desired benzamide product was isolated (Scheme 4.4).

The Tiffeneau–Demjanov rearrangement has been used as the initial step in the flow synthesis of 1-amino-2,5-anhydro-D-mannose (“mannitolamine”)—a key intermediate for the synthesis of various fluorophores, conjugates, or photoaffinity labels [9]. Starting from glucosamine hydrochloride, the synthesis of mannitolamine was achieved in three subsequent flow steps—rearrangement, oxime formation, and hydrogenolysis. The first two steps were performed in an uninterrupted fashion—at 100 °C and 40 °C correspondingly. However, for the final hydrogenolysis step, a simple workup was required to remove the water-soluble salts and by-products (Scheme 4.5). The hydrogenolysis was successfully performed at 70 °C on a 10 % Pd/C catalyst using 100 bar of hydrogen and 1 ml/min flow rate.

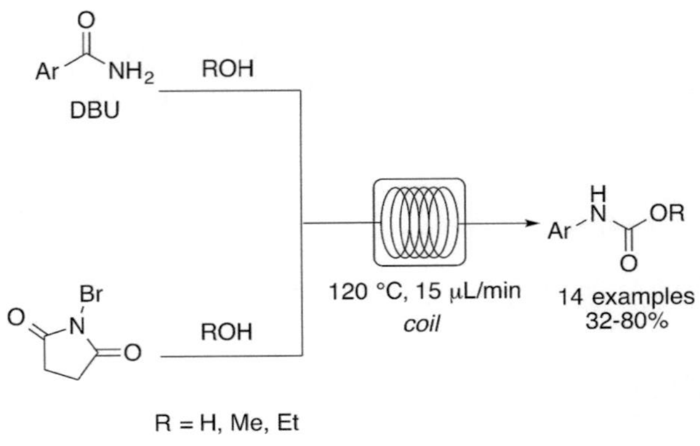

Scheme 4.3 Continuous flow Hofmann rearrangement

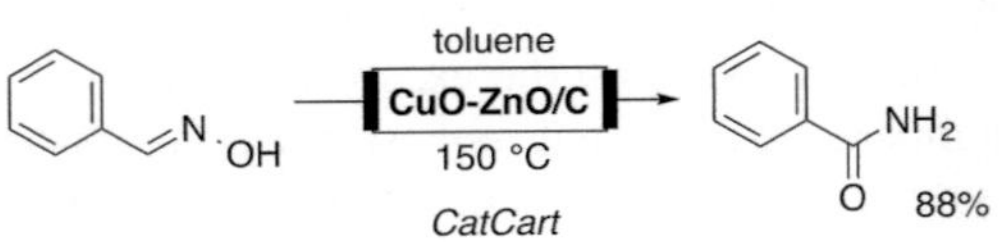

Scheme 4.4 Cuo/ZnO-catalyzed Beckmann rearrangement

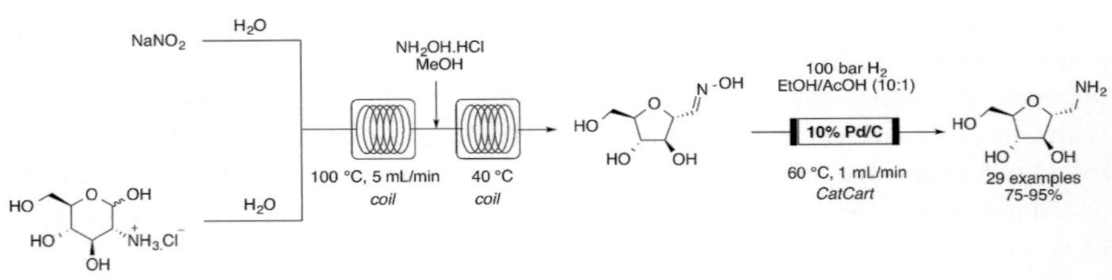

Scheme 4.5 Mannitolamine synthesis via the Tiffeneau–Demjanov rearrangement

References

1. M. Baumann, I.R. Baxendale, S.V. Ley, N. Nikbin, C.D. Smith, J.P. Tierney, Org. Biomol. Chem. **6**, 1577 (2008)
2. M. Baumann, I.R. Baxendale, S.V. Ley, N. Nikbin, C.D. Smith, Org. Biomol. Chem. **5**, 1587 (2008)
3. L. Guetzoyan, R.J. Ingham, N. Nikbin, J. Rossignol, M. Wolling, M. Baumert, N.A. Burgess-Brown, C.M. Strain-Damerell, L. Shrestha, P.E. Brennan, O. Fedorov, S. Knapp, S.V. Ley, Chem. Med. Chem. **5**, 540 (2014)
4. M. Mohamed, T.P. Goncalves, R.J. Whitby, H.F. Sneddon, D.C. Harrowven, Chem. Eur. J. **17**, 13698 (2011)
5. E. Packard, D.D. Pascoe, J. Maddaluno, T.P. Goncalves, D.C. Harrowven, Angew. Chem. Int. Ed. **52**, 13706 (2013)

6. J.T. Hill-Cousins, I.-A. Pop, G. Pileio, G. Stevanato, P. Hakansson, S.S. Roy, M.H. Levitt, L.J. Brown, R.C.D. Brown, Org. Lett. **17**, 2150 (2015)
7. A. Palmieri, S.V. Ley, K. Hammond, A. Polyzos, I.R. Baxendale, Tetrahedron Lett. **50**, 3287 (2009)
8. S.K. Sharma, S.D. Bishopp, C.L. Allen, R. Lawrence, M.J. Bamford, A.A. Lapkin, P. Plucinski, R.J. Watson, J.M.J. Williams, Tetrahedron Lett. **52**, 4252 (2011)
9. M.B. Plutschack, D.T. McQuade, G. Valenti, P.H. Seeberger, Beilstein J. Org. Chem. **9**, 2022 (2013)

Chapter 5
Organic Synthesis in Dedicated Continuous Flow Systems

Cycloaddition Reactions

5.1 Diels–Alder Cycloadditions

Cycloaddition reactions are among the most often-studied transformations using continuous flow technology, and a plethora of examples can be found in the literature.

The first example of a cycloaddition reaction in a commercial device—an Ehrfeld reactor—was the Diels–Alder addition of maleic anhydride with 2,3-dimethylbuta-1,3-diene. The resulting product—3a,4,7,7a-tetrahydro-5,6-dimethylisobenzpfuran-1,3-dione—was obtained on a gram scale in 98 % yield after only 30 min reaction time at 60 °C [1].

A high-temperature, high-pressure Diels–Alder reaction was also demonstrated on an X-Flash instrument to illustrate the capabilities of the flow equipment in direct comparison with a microwave batch reactor [2]. An electrospray mass spectrometer has been used to monitor the Diels–Alder continuous flow synthesis of 1,4-endioxide-1,4-dihydronaphtalene [3].

By varying the reaction conditions—temperature, flow rate, and reagent equivalents—the Diels–Alder functionalization of C_{60} or C_{70} fullerene was achieved in a continuous flow protocol [4]. Indene was reacted with the corresponding fullerenes at 220 °C, providing access to the bisadducts. Twenty to thirty-six equivalents of indene were applied and reacted at superheated conditions. In less than 2 h, 54 % of the C_{60} bisadduct and 49 % of the C_{70} bisadduct were obtained. The same fullerenes were successfully subjected to a [1 + 3]-cycloaddition with a tosylhydrazone precursor.

Intermolecular hetero-Diels–Alder cycloaddition reactions of substituted nitrosodienophiles and a set of selected 1,3-dienes have been evaluated by the research group of Stevens (Scheme 5.1) [5]. Numerous reaction parameters were varied to determine the optimal reaction conditions—temperature (0–95 °C), pressure (1–100 bar), and flow rate (0.2–1 mL/min)—as well as several different solvents (acetone, methanol, THF, acetonitrile, and DMF). While in few of the

T. Glasnov, *Continuous-Flow Chemistry in the Research Laboratory*,
DOI 10.1007/978-3-319-32196-7_5

Scheme 5.1 Continuous flow hetero-Diels–Alder cycloaddition

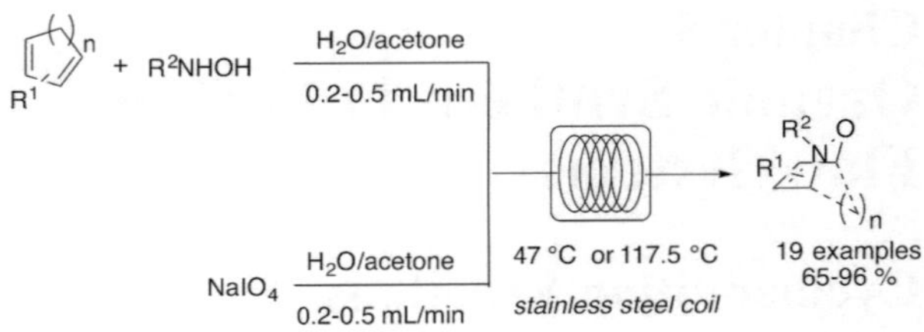

studied examples no products were obtained even at elevated temperatures, it was possible to collect moderate to good yields of cycloadducts in the most cases. The corresponding 1,2-oxazines could be generated in continuous manner and on a 0.2–1 g/h scale.

An intramolecular inverse electron demand hetero-Diels–Alder reaction of substituted pyrimidines led to the formation of bicyclic annulated pyridines [6]. Typically, such transformations require rather high reaction temperatures and extended reaction times. Respectively, high-boiling solvents, which are often considerably toxic, have to be used. Furthermore, stoichiometric amounts of hydrogen cyanide are generated, making the overall process highly hazardous when performed in a conventional fashion. In contrast, working in the contained environment of a flow reactor, many of the related hazards can be eliminated. The formation of dihydro-5*H*-[1]-pyridine was carried out at 250 °C in toluene, containing 1 % v/v pentan-3-one to avoid blockage of the reactor coil and to scavenge the formed HCN as cyanohydrin adduct. The running process was stable for several hours and delivered 21 g (84 %) of the desired product (Scheme 5.2). An extended library was then generated using an in-house build device, able to work at even higher temperatures.

In another example, a heteropoly acid-catalyzed three-component aza-Diels–Alder synthesis of azabicyclo-[2.2.2]-octan-5-ones was successfully achieved [7]. Phosphotungstic acid was used as the catalyst of choice and provided up to 98 % conversion with a 7:93 ratio of endo-/exoproducts. Using an inline water-absorption module, optimal catalytic activity and hence maximum conversion were ensured. With precise temperature and mixing control, high conversions at 120 °C and 8 min residence time were possible.

5.2 [3 + 2] Cycloadditions

Azides are the key intermediates for the synthesis of triazoles by the copper-catalyzed azide–alkyne cycloaddition reaction (CuAAC). Typically, immobilized copper(I) sources are used in such a process. An intriguing example of a 1,2,3-triazole synthesis was presented by Ley and coworkers [8]. Starting from aldehydes or even alcohols and azides, the corresponding triazoles were prepared by a

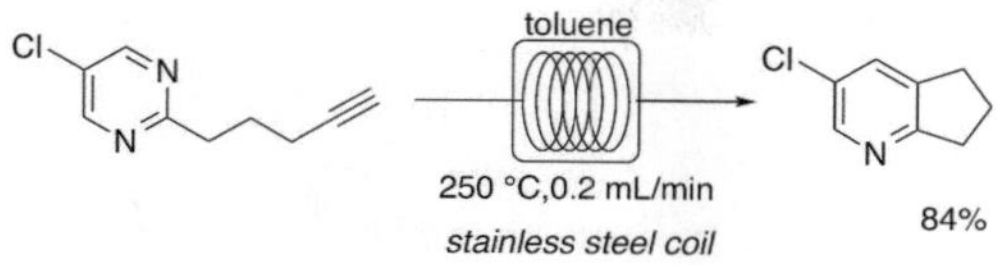

Scheme 5.2 An intramolecular inverse electron demand hetero-Diels–Alder process

continuous flow reaction combined with an in-line purification sequence. The Bestmann–Ohira reagent was used to convert the starting aldehydes into the respective alkynes upon the Seyferth–Gilbert homologation within 30 min residence time at temperature of 100 °C in a 10 mL PFA tube reactor. The reaction stream, containing the alkyne and the azide, was then directed over a reactor segment containing immobilized CuI as a catalyst, for the subsequent [3 + 2] conversion of the alkynes into triazoles (Scheme 5.3). The product stream was directly purified via series of in-line scavenger cartridges to provide the pure triazoles.

A group at Wyeth Research developed a continuous process to satisfy the company's needs for *N*-alkylated 5-amino-1,2,3-triazole carboxamides [9]. Unfortunately, the considerable safety hazards in connection with small molecular weight alkyl azides needed for the envisaged synthesis precluded their direct use. To resolve these issues, a continuous flow approach was sought. Additionally, more stable azide building blocks, containing a large sulfur substituent, were employed as starting materials. These substituents can be easily removed by a RaneyNi desulfurization. β-Azidomethyl phenyl sulfide was prepared as an appropriate starting material for this study, and its thermal properties were evaluated. A subsequent continuous flow [3 + 2] cycloaddition experiment with cyanoacetamide delivered the expected cycloadduct in 90 % isolated yield after only 2 min residence time at 65 °C. A RaneyNi desulfurization successfully concluded the synthesis with a 90 % yield of the final triazole carboxamide. With the aim of preparing various analogues of Brilinta®—an antiplatelet agent—a library of *N*-substituted 5-amino-1,2,3-triazole carboxamides was synthesized by Jones et al. [10]. In this report, the required azide was generated in situ by using sodium azide in combination with various chlorides or bromides. Instant reaction with cyanoacetamide as above furnished the desired cycloadducts (Scheme 5.4). The subsequent batch functionalization provided triazolopyrimidine analogues of Brilinta®.

Copper powder, packed in a cartridge, was used as a readily accessible Cu (I) source to promote the synthesis of various 1,2,3-triazole-modified β-aminocyclohexanecarboxylic acid derivatives via the CuAAC reaction [11]. Depending on the type and quantity of additives—an organic base or acid—different extent of leaching was observed. Nevertheless, 12 highly functionalized products were obtained at temperatures between ambient and 100 °C and 100 bar pressure in dichloromethane as a solvent and 0.5 mL/min flow rate (Scheme 5.5).

The same group employed this approach for the synthesis of highly functionalized cispentacin derivatives as well [12]. Considering the use of immobilized copper(I) salts as catalyst in the CuAAC to generate triazoles, a study on the applicability of copper on charcoal for continuous flow processing

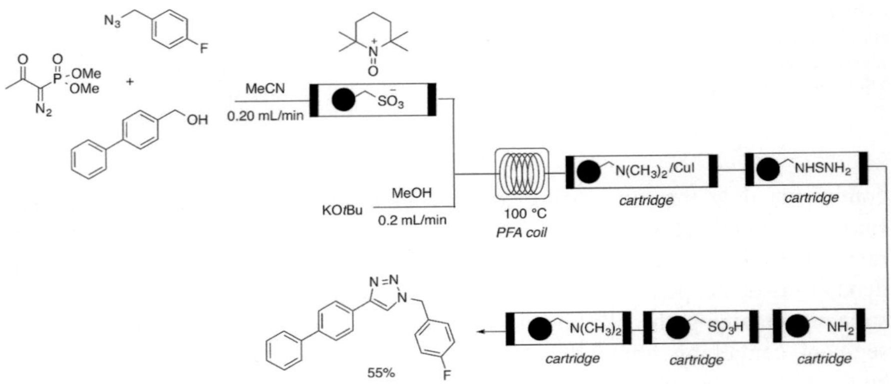

Scheme 5.3 Copper-catalyzed azide–alkyne cycloaddition under flow conditions

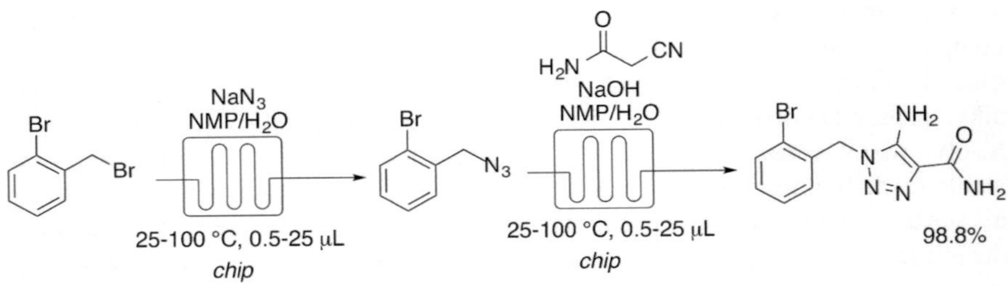

Scheme 5.4 Synthesis of *N*-substituted 5-amino-1,2,3-triazole carboxamides as precursors of Brilinta® analogues

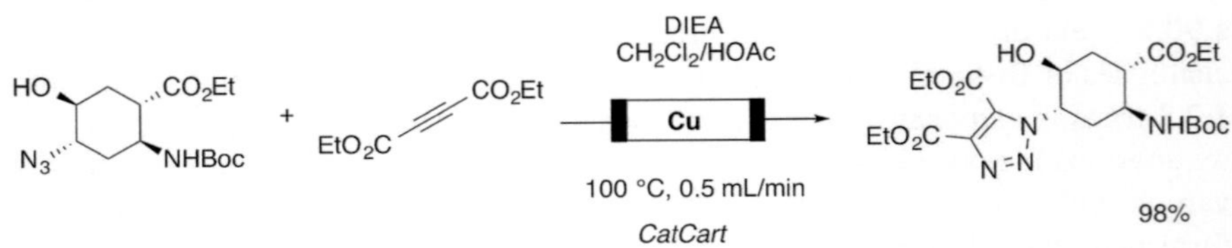

Scheme 5.5 A 1,2,3-triazole-modified β-aminocyclohexanecarboxylic acid derivative synthesis via the CuAAC reaction

was recently carried out. Confirming the homogeneous mechanism of the cycloaddition, a significant Cu leaching from the packed-bed heterogeneous catalyst was observed [13]. The copper can usually be removed from the product stream in-line with the aid of metal scavenger cartridges. The latter is a powerful purification technique, but both the catalyst and the scavenger cartridges have to be periodically replaced, thus interrupting the flow process. Therefore, CuAAC reactions with homogeneous copper catalysts are often preferred. To address such issues, Bogdan and James recently presented an interesting concept [14]. Successful intramolecular

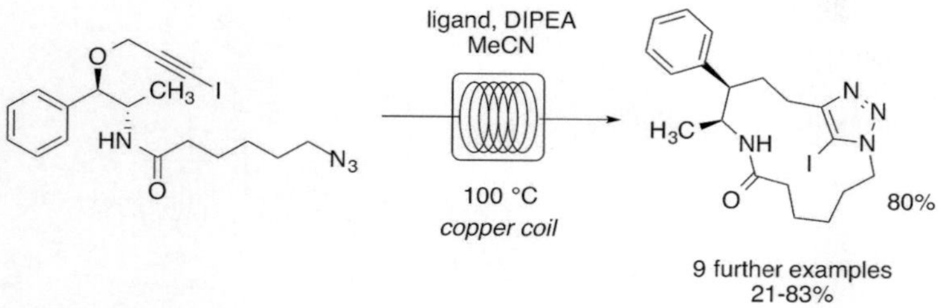

Scheme 5.6 Macrocycle synthesis using copper tube reactor

cycloaddition was performed in a copper coil as both reactor and catalyst source at temperatures of 150 °C in DMF/water mixture as a solvent. A variety of 12- to 31-membered triazole-containing rings were obtained after short residence times of 5–10 min (22–90 % yield) (Scheme 5.6).

A similar approach was considered for the assembling of macrocyclic peptoids in a sequential Ugi/CuAAC flow synthesis [15]. This complex process required significant fine-tuning and optimization until an isocyanide generation, an azide generation, and a Ugi four-component process could be combined into a single continuous operation to prepare the intermediate linear peptoids, containing an azide and an alkyne moiety. These were then subjected to a final CuAAC-macrocyclization into a copper tube reactor at 140 °C for 25 min residence time to obtain preferably monomeric cyclopeptoids in good yields.

As an alternative method for the in situ synthesis of organic azides, the use of stable and nonexplosive reagents—such as *tert*-butylnitrite (t-BuONO) and azidotrimethylsilane ($TMSN_3$) in dry acetonitrile as solvent—was also considered recently for the production of *N*-aryl-1,2,3-triazoles [16]. The synthesis has however a thermal hazard potential, and for this reason, a continuous flow processing was envisaged. As in previous examples, the in situ-prepared azide was immediately reacted with a β-ketoester in the presence of a base to form the 1,2,3-triazoles. The first step, the reaction of an aromatic amine, $TMSN_3$, and *t*-BuONO in acetonitrile at 50 °C and 20–30 min residence time, was directly followed by the triazole formation upon the addition of β-ketoester and DBU as organic base in the second step, taking place at 80 °C with residence time of 13–19 min. The N-aryl-1,2,3-triazoles were isolated in 54–79 % yield after a NH_4Cl-quench, followed by extraction and chromatography (Scheme 5.7).

Tetrazoles can be synthesized in a similar fashion as triazoles applying the Huisgen 1,3-dipolar cycloaddition of azides to nitriles. Tetrazole chemistry has attracted much attention in the past years, which may be due in part to the role of the tetrazole heterocycle as a metabolically stable alternative of the carboxylic acid functionality in pharmaceutically active agents such as Losartan, Valsartan, Candesartan, Irbesartan, and Olmesartan. A high-temperature continuous flow process for the synthesis of tetrazoles, making use of the 1,3-diploar cycloaddition of hydrazoic acid to nitriles, was reported in 2010 and reevaluated in 2012 [17–19]. The volatile and explosive hydrazoic acid was generated in situ by proper

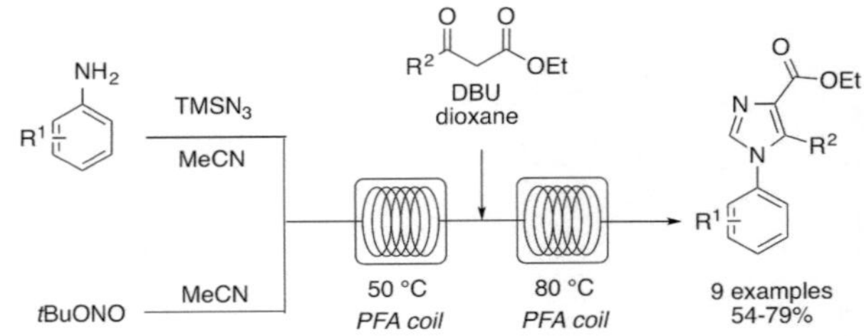

Scheme 5.7 Alternative in-line generation and cycloaddition of azides under continuous flow conditions

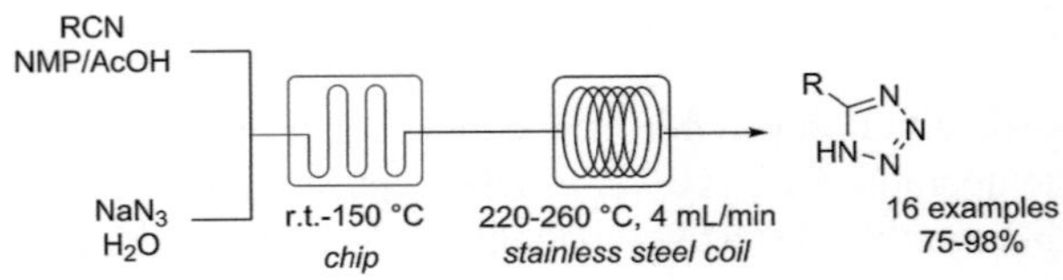

Scheme 5.8 High-temperature 1,3-dipolar cycloaddition of hydrazoic acid to nitriles

mixing of aqueous sodium azide and acetic acid in a static glass mixer. The generated HN_3 was directly consumed by the cycloaddition to nitriles in an additional heated coil reactor (Scheme 5.8). The effluent product stream was passed through a heat exchanger before the unconsumed hydrazoic acid was destroyed with aqueous $NaNO_2$. The cycloadditions were typically carried out at temperatures of 220 °C requiring 10–15 min to achieve full conversion. Depending on the used substrates, yields in the range of 75–98 % were obtained. The Jamison group described very similar tetrazole synthesis in the following year, whereby the quench of the excess HN_3 was also accomplished in-line [20].

An alternative [3 + 2] cycloaddition reaction was used to prepare 1,2,4-tetrazoles [21]. A flow procedure was developed to form directly ethylisocyanoacetate from *N*-formylglycine with the help of triphosgene. In a second stream, various diarylazonium species were readily synthesized from the corresponding anilines or their mono-hydrochloride salts upon treatment with *t*-BuONO. The two convergent streams were then diluted with ethanol and aq. K_2CO_3 before heating at 75 °C for 52 min (Scheme 5.9). The products were isolated in high purity and yields using a standard workup procedure.

Ley and coworkers demonstrated another [3 + 2] cycloaddition process involving in situ generation of unstabilized azomethine ylides [22]. The synthesis provided access to numerous 3-nitropyrrolidines. The prepared heterocycles are useful building blocks for several API synthesis since pyrrolidine-based compounds from synthetic and natural origin have shown various interesting biological effects. A small collection of 3-nitropyrrolidines was prepared by mixing two reagent streams—one with a corresponding nitroalkene and a second one containing *N*-(methoxymethyl)-*N*-(trimethylsilyl)benzylamine in acetonitrile or toluene. Depending on the substrate, mixing and reacting of the two streams at 60–120 °C for 30–90 min provided the desired products in 74–93 % yield (Scheme 5.10). TFA acid or a fluoride monolith generate the reactive dipole for the cycloaddition

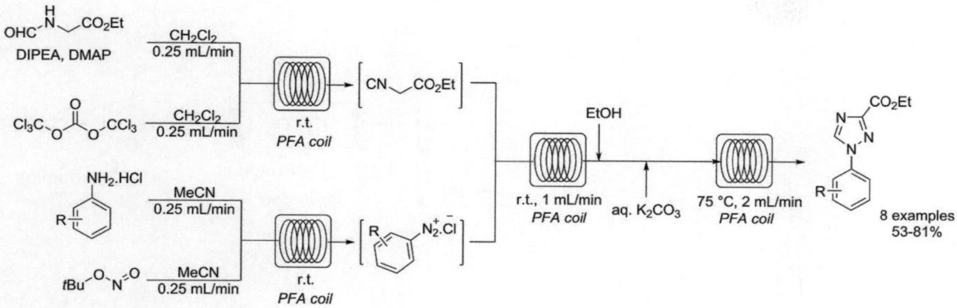

Scheme 5.9 Multistage synthesis of 1,2,4-traizoles via [3 + 2] cycloaddition reaction

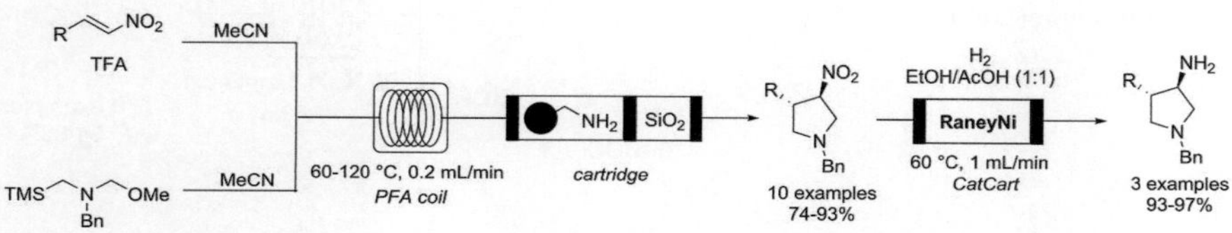

Scheme 5.10 Generation of unstabilized azomethine ylides for the synthesis of 3-nitro- and 3-aminopyrrolidines

reaction. Subsequent nitro-reduction was also achieved in a continuous flow reactor—an H-Cube hydrogenator—to produce 3-aminopyrrolidines. An identical flow process was also reported by the group of Fray [23].

5.3 Miscellaneous Cycloadditions

A comparison between conventional batch, microwave batch, and flow-assisted methodologies for the synthesis of Δ^2-isoxazolines was disclosed by Conti et al. [24]. The [1 + 3] cycloaddition of nitrile oxides and alkenes was used as a convenient one-step approach for this synthesis. Initially, a rapid screening of base type, reaction temperature, flow rate, and stoichiometry was performed. At 90 °C temperature, only 10 min reaction time and using K_2CO_3 as a solid base packed in a glass cartridge, satisfactory yields of the target cycloadducts were obtained from a mixture of the corresponding chloroximes and *N*-Boc-Δ^3-pyrroline (Scheme 5.11). Higher temperatures had detrimental effect. Under conventional conditions, the reaction required several days to provide <45 % yield, while in the microwave only 1.5 h was sufficient for the synthesis. The flow approach reduced the reaction time even further to only 10 min while simultaneously increasing the yields. The chloroxime approach was also applied for the synthesis of kainic acid analogues in the search for potent agonists of the kainic receptors in neurons which affect neuronal signaling in the treatment of disorders like chronic pain, stroke, epilepsy, etc. [25].

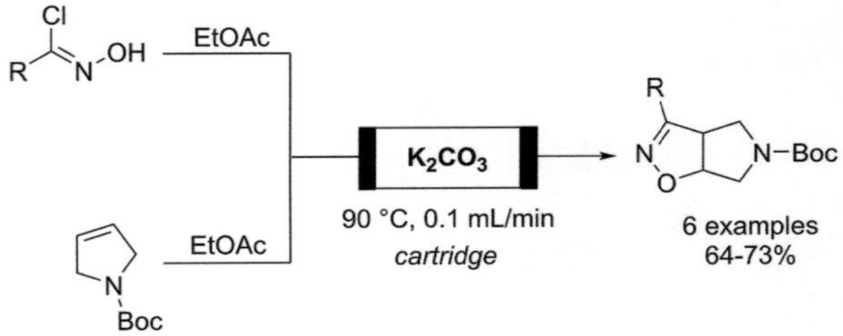

Scheme 5.11 Synthesis of Δ^2-isoxazolines from chloroximes and *N*-Boc-Δ^3-pyrroline

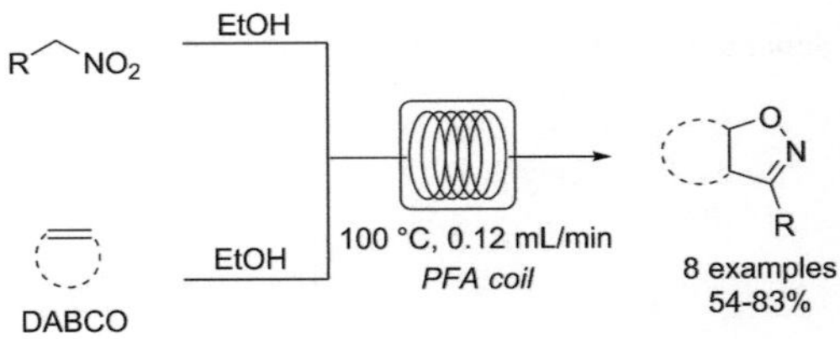

Scheme 5.12 Mono- and bicyclic isoxazoles synthesis via a cycloaddition reaction

A similar method was deployed for the generation of mono- and bicyclic isoxazoles. Dehydrogenation of terminal nitrocompounds was used to generate the required nitrile oxides for the subsequent cycloaddition with various alkenes [26]. Two approaches were tested to acquire the nitrile oxides—either in the presence of di-*t*-butyl dicarbonate (Boc_2O) and 4-dimethylaminopyridine (DMAP) or just in the presence of another organic base—1,4-diazabicyclo[2.2.2] octane (DABCO) (Scheme 5.12).

The Boc_2O method was applied for unactivated nitroalkanes while the DABCO method for activated substrates. The Boc_2O approach required only 50 °C to provide the products within 40 min reaction time, while the DABCO approach worked at 100 °C and required a reaction time of 250 min. The better mixing properties can possibly serve as an explanation for the observed shorter reaction times as compared to corresponding batch experiments. Nevertheless, both methods provided comparable yields.

References

1. D.A. Snyder, C. Noti, P.S. Seeberger, Helv. Chim. Acta **88**, 1 (2005)
2. T. Razzaq, T.N. Glasnov, C.O. Kappe, Eur. J. Org. Chem. 1321 (2009)
3. D.L. Browne, S. Wright, B.J. Deadman, S. Dunnage, I.R. Baxendale, R.M. Turner, S.V. Ley, Rapid Commun. Mass Spectrom. **26**, 1999 (2012)
4. H. Seyler, W.W.H. Wong, D.J. Jones, A.B. Holmes, J. Org. Chem. **76**, 3551 (2011)
5. J.-C.M.R. Monbaliu, A. Cukalovic, J. Marchand-Brynaert, C.V. Stevens, Tetrahedron Lett. **51**, 5830 (2010)
6. R.E. Martin, F. Morawitz, C. Kuratli, A.M. Alker, A.I. Alanine, Eur. J. Org. Chem. 47 (2012)
7. W. He, Z. Fang, Z. Yang, D. Ji, K. Guo, RSC Adv. **5**, 58798 (2015)
8. I.R. Baxendale, S.V. Ley, A.C. Mansfield, C.D. Smith, Angew. Chem. Int. Ed. **48**, 4017 (2009)

9. R. Tinder, R. Farr, R. Heid, R. Zhao, R.S. Rarig Jr., T. Storz, Org. Process Res. Dev. **13**, 1401 (2009)
10. S. Sadler, M.M. Sebeika, N.L. Kern, D.E. Bell, C.A. Laverack, D.J. Wilkins, A.R. Moeller, B.C. Nicolaysen, P.N. Kozlowski1, C. Wiles, R.J. Tinder, G.B. Jones, J. Flow Chem. **4**, 140 (2014)
11. S.B. Ötvös, A. Georgiades, I.M. Mandity, L. Kiss, F. Fülöp, Beilstein J. Org. Chem. **9**, 1508 (2013)
12. S.B. Ötvös, I.M. Mandity, L. Kiss, F. Fülöp, Chem. Asian J. **8**, 800 (2013)
13. M. Fuchs, W. Goessler, C. Pilger, C.O. Kappe, Adv. Synth. Catal. **352**, 323 (2010)
14. A.R. Bogdan, K. James, Org. Lett. **13**, 4060 (2011)
15. C.E.M. Salvador, B. Pieber, P.M. Neu, A. Torvisco, C.K.Z. Andrade, C.O. Kappe, J. Org. Chem. **80**, 4590 (2015)
16. F. Stazi, D. Cancogni, L. Turco, P. Westerduin, S. Bacchi, Tetrahedron Lett. **51**, 5385 (2010)
17. B. Gutmann, J.-P. Roduit, D. Roberge, C.O. Kappe, Angew. Chem. Int. Ed. **49**, 7101 (2010)
18. B. Gutmann, D. Obermayer, J.-P. Roduit, D.M. Roberge, C.O. Kappe, J. Flow Chem. **1**, 8 (2012)
19. B. Gutmann, T.N. Glasnov, T. Razzaq, W. Goessler, D.M. Roberge, C.O. Kappe, Beilstein J. Org. Chem. **7**, 503 (2011)
20. P.B. Palde, T.F. Jamison, Angew. Chem. Int. Ed. **50**, 3525 (2011)
21. M. Baumann, A.M.R. Garciaa, I.R. Baxendale, Org. Biomol. Chem. **13**, 4231 (2015)
22. M. Baumann, I.R. Baxendale, S.V. Ley, Synlett. **21**, 749 (2010)
23. M. Grafton, A.C. Mansfield, M.J. Fray, Tetrahedron Lett. **51**, 1026 (2010)
24. S. Castellano, L. Tamborini, M. Viviano, A. Pinto, G. Sbardella, P. Conti, J. Org. Chem. **75**, 7439 (2010)
25. L. Tamborini, F. Mastronardi, G. Cullia, R. Ettari, C. De Micheli, ARKIVOC **iv**, 377 (2013)
26. M. Brasholz, S. Saubern, G.P. Savage, Aust. J. Chem. **64**, 1397 (2011)

Chapter 6
Organic Synthesis in Dedicated Continuous Flow Systems

Reductions and Oxidations

6.1 Reductions

6.1.1 Hydrogenation

After the first publication back in 2005 and the introduction on the market [1, 2], the H-Cube flow hydrogenation system has found widespread application in synthetic laboratories around the world. In part, the success of this reactor concept surely can be associated with the hassle of handling bottled hydrogen gas which has been always associated with severe precaution measures and is typically performed in specially designed and equipped "hydrogenation" rooms—problems that the H-Cube elegantly eliminates. Using an electrolysis cell, hydrogen is produced on demand and in sufficient quantities and pressures up to 100 bar. The required catalysts are prepacked in exchangeable, ready-for-use cartridges. Reduction of various chemical species is easily achievable at low costs and efforts [1–8].

Different precious metal catalysts and complexes can be used for the hydrogenation process. Several studies exist that are dealing with the evaluation of performance, stability, and effectiveness of different solid supports for anchoring these catalysts [9–13]. Cinchonine and quinidine complexes of platinum were tested as chiral catalysts for the flow reduction of ethyl pyruvate, methyl benzoylformate, and 2,2,2-trifluoroacetophenone [14–19].

On the route to spirocyclic piperidines, Arnott and coworkers applied a flow hydrogenation to prepare the desired products [20]. The valuable spirocyclic moieties contain a fully saturated piperidine ring that is derived from dihydropyridines. The process was successfully performed at 100 bar hydrogen pressure and 60 °C reaction temperature, using Pd/C as a catalyst and ethanol/ethyl acetate as solvent mixture. A spirocyclic piperidine, structurally related to the biologically active molecule MK677, was also prepared using the optimized hydrogenation conditions (Scheme 6.1).

T. Glasnov, *Continuous-Flow Chemistry in the Research Laboratory*,
DOI 10.1007/978-3-319-32196-7_6

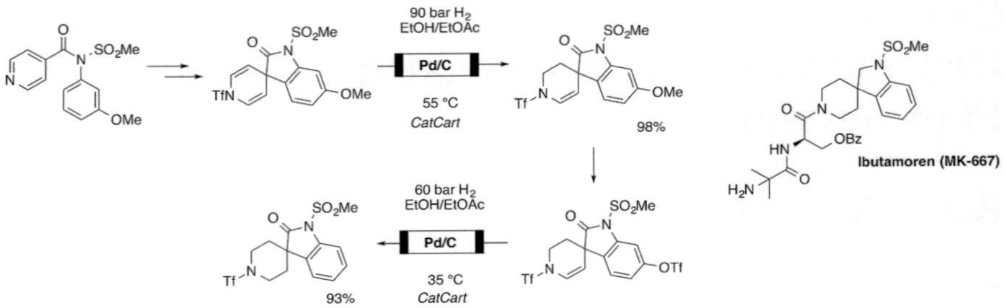

Scheme 6.1 Continuous flow hydrogenation in the synthesis of spirocyclic piperidines as structural analogues of Ibutamoren

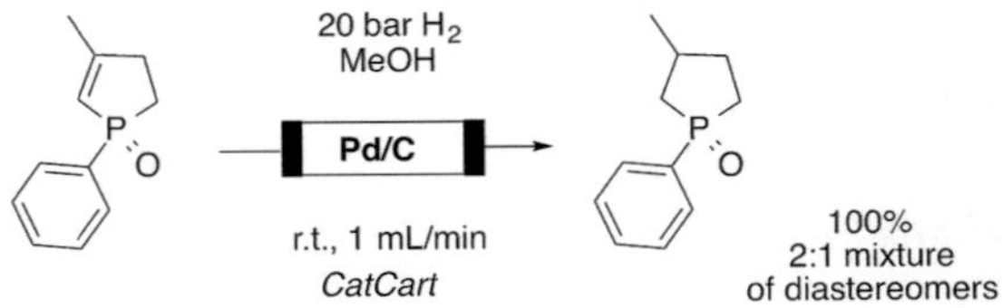

Scheme 6.2 Synthesis of phosphine oxide precatalyst via flow hydrogenation

In another study, functionalized pyridines could also be conveniently prepared under flow hydrogenation conditions [21]. To avoid solubility issues and to increase the productivity, the reaction had to be performed in water or acetic acid as solvent. Furthermore, the possibility to generate deuterated compounds was also successfully demonstrated—the water reservoir of the H-Cube instrument had to be filled simply with commercially available heavy water. Similar studies extended the deuteration strategy to further organic species beyond cinnamic acid [22, 23]. An interesting application example is also the partial deuteration of alkynes, particularly for the preparation of a deuterated taxol side chain, as reported by Chandrasekhar and coworkers [24]. Potentially bioactive chalcone derivatives were also synthesized using the described safe and time- and cost-effective continuous flow deuteration strategy [25].

In a study to develop a catalytic Wittig process, O'Brien et al. employed 3-methyl-1-phenylphospholane-1-oxide as more suitable reagent instead of the traditionally used triphenylphosphine oxide [26]. The phosphine oxide precatalyst was prepared relatively easy by a double-bond reduction of the corresponding starting material over Pd/C in methanol and 20 bar hydrogen pressure (Scheme 6.2).

Several Tröger's base analogues were recently synthesized [27]. Seeking access to the corresponding diamino derivatives, two different reduction conditions were tested by Try et al. While the application of tin(II) chloride proved to be unsuccessful, a relatively simple nitro-reduction under flow conditions with hydrogen gas over Pd/C provided the desired diamines in 85 and 97 % yields after simple solvent removal (Scheme 6.3).

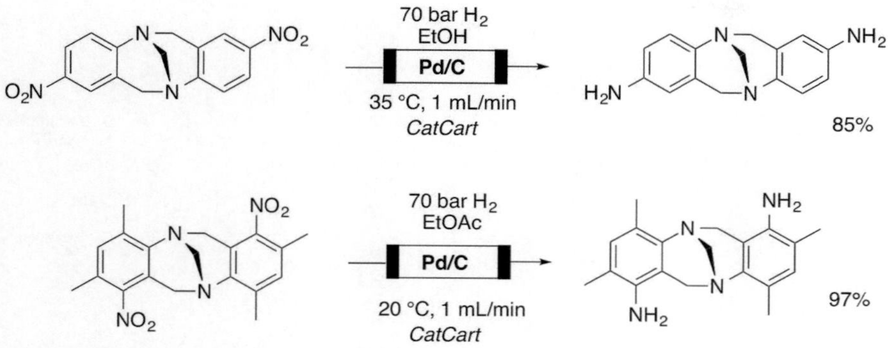

Scheme 6.3 Nitro-reduction in the synthesis of Type I and II diamines

A selective nitro-reduction was also used for the continuous synthesis of Boscalid®. In the presence of halogen substituents, the use of Pt/C was preferential, since Pd/C did not provide enough selectivity for the reduction process [28]. The synthesis of pyrido[2,3-d]pyrimidin-5-ones as a novel class of anti-inflammatory macrophage colony-stimulating factor-1 receptor inhibitors involved a flow nitro-reduction step as well [29]. In the search for novel proteasome inhibitors as cancer cell inhibitors, a library of naphthoquinones has been synthesized [30]. In order to introduce an additional functionality, various nitro-compounds were subjected to flow hydrogenation to obtain the corresponding amines. In the same manner, nitro-reduction was used as a key step also in the preparation of 4-(pyrazol-1-yl) carboxanilides useful as inhibitors of canonical transient potential receptor channels [31]. The reduction was performed either as microwave batch transfer hydrogenation or as a flow hydrogenation. The best results were obtained with Pd/Al_2O_3 as catalyst for the flow process. With the help of an H-Cube instrument, the patented five-step synthesis of SEA0400—a selective inhibitor of the Na+/Ca2+ exchanger—was shortened to three steps with an overall yield of 52 % [32]. A flow nitro-reduction of ^{18}F-labeled nitrobenzene conveniently afforded the desired aniline, while a flow debenzylation protocol provided [^{18}F]-CABS13 [33]. As a part of an industrial cooperation project, an automated library synthesis of various AbI kinase inhibitors was achieved [34]. Besides the impressive automation features of the overall process, a nitro-reduction was implemented as a part of the process. Nitro-reduction was used also as an intermediate step in the synthesis of a potent $5HT_{1B}$ antagonist [35]. The same reaction type was used for the preparation of various heterocycles—riboflavins, quinoxalines, and benzodiazepines [36]. Processing of substituted 1,2-nitroanilines through an H-Cube hydrogenator resulted in the synthesis of 1,2-dianilines that were subsequently reacted with alloxane without isolation to provide various riboflavins. Furthermore, the reduction of several 2-aminobenzophenones over PtO_2 in CH_2Cl_2/methanol mixture at 60 °C directly allowed the isolation of the relevant dibenzodiazepines in only 5 min reaction time. In-line cyclodehydration was aided by a cartridge of anhydrous $MgSO_4$

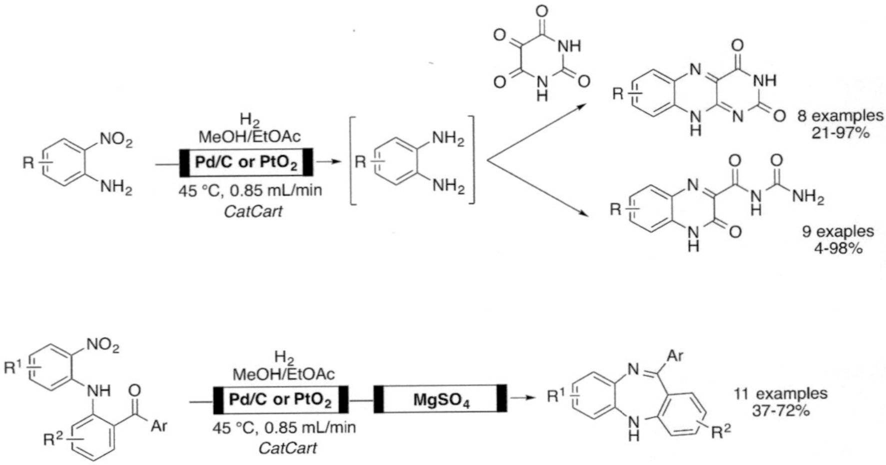

Scheme 6.4 Synthesis of various heterocycles involving flow nitro-reduction

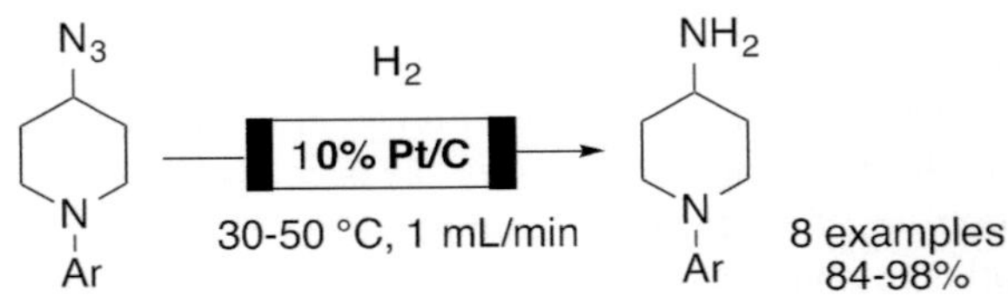

Scheme 6.5 Synthesis of 1-aryl-4-aminopiperidines via continuous flow azide reduction

(Scheme 6.4). Recently, an almost identical approach was employed to successfully prepare 1,4-benzodiazepine-5-ones from corresponding benzamides [37].

Besides nitro-reduction, the hydrogenation of azides is a popular alternative pathway to prepare amines. Using the H-Cube instrument, a combination of methanol as a solvent, Pd/C as catalyst, and atmospheric pressure of hydrogen at room temperature, several aliphatic azides were easily transformed into the corresponding amines [38]. Addition of di-*tert*-butyl dicarbonate (Boc_2O) was required to prevent acetyl group migration whenever azido acrylamides were used as starting materials. A mild azide reduction process was reported by a group at Amgen [39]. A library of 1-aryl-4-aminopiperidines was synthesized. The mild conditions selectively converted the azide group into an amine, leaving the present nitrile substituent intact (Scheme 6.5).

In a "one-pot" procedure, acid ester or acid amide derivatives of norcantharidin were conveniently prepared [40]. The standard batch-wise synthesis is traditionally sluggish due to the resistance of 5,6-dehydronorcantharidin to hydrogenation. This was not the case when performing the reaction under flow conditions. At 50 °C reaction temperature and 50 bar of hydrogen, sufficient quantities of norcantharidin and analogues were obtained for subsequent biological screenings (Scheme 6.6).

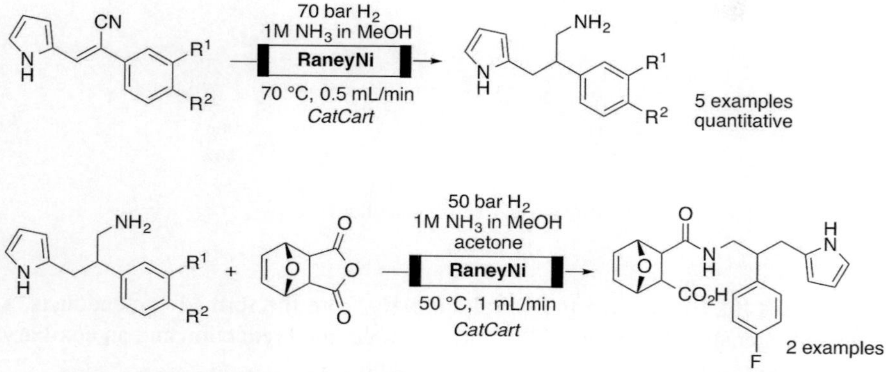

Scheme 6.6 Acid-amide derivatives of norcantharidin obtained under continuous flow hydrogenation conditions

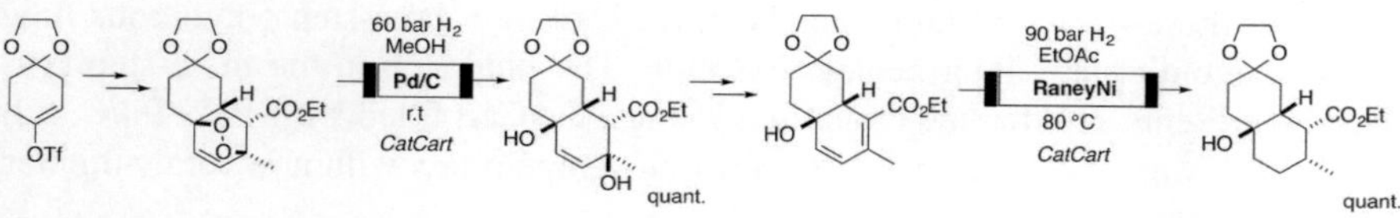

Scheme 6.7 Continuous flow hydrogenation towards the synthesis of a natural product

A number of 2-phenyl-3-(1*H*-pyrrol-2-yl)propan-1-amines were prepared and reacted with 5,6-dehydronorcantharidin to yield the corresponding ring-opened acid amides, which are promising protein phosphatase 1 and 2A inhibitors [41]. Similarly, a nitrile-to-amine reduction was utilized to diversify a small library of substituted oxazoles [42]. Further studies devoted to the generation of highly decorated norcantharidin analogues utilized selective reductive aminations, nitrile, and olefin reductions [43]. Besides the required reaction optimization studies, few catalysts were examined for the different reductive syntheses—Pd/C, $Pd(OH)_2/C$, Pt/C, RaNickel, etc.

In the synthesis of nabumetone, an anti-inflammatory drug, a two-step flow process was realized. As the second and last synthetic step, selective double-bond flow hydrogenation in the presence of a carbonyl group was possible. Using RaneyNi as the catalyst of choice, reaction temperature of 100 °C and atmospheric pressure of hydrogen, full substrate conversion, and 90 % yield were reported [44].

Flow hydrogenation at elevated temperatures was highly rewarding for the synthesis of some natural product analogues (crotonins) where other popular batch methods failed [45]. Quantitative double-bond hydrogenation was possible at a temperature of 80 °C and 90 bar hydrogen pressure to prepare the 5-β-hydroxy-*cis*-dehydrocrotonin core (Scheme 6.7). The substrate for this process was also synthesized using a flow double-bond hydrogenation and peroxide ring opening.

Another study involving the diastereoselective 1,3-cycloadditon of pyrylium ylides with chiral enamides investigated the use of chiral oxazolidinones as chiral

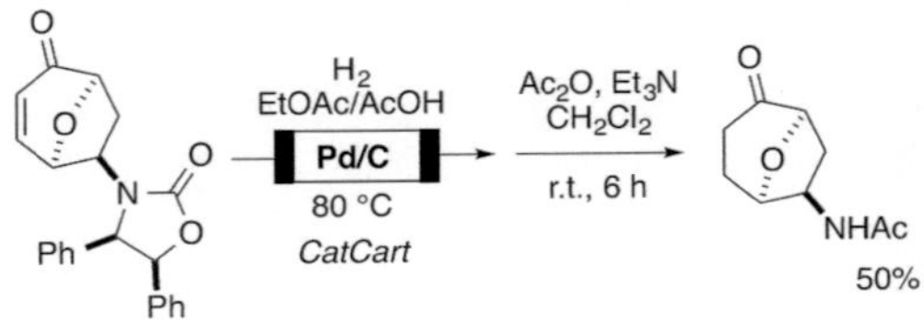

Scheme 6.8 Chiral auxiliary cleavage under flow conditions

auxiliaries [46]. To remove the chiral auxiliary from the formed cycloadducts, a flow hydrogenation was used where both a double-bond reduction and an auxiliary cleavage were simultaneously achieved (Scheme 6.8).

The retrosynthetic analysis of atazanavir, an antiretroviral drug approved by the US Food and Drug Administration for HIV treatment, reveals three different building blocks required for the assembly of the active molecule. The synthesis of one of these—a biaryl unit—has been realized in a three-step continuous flow process, avoiding any intermediate isolation. The final step in the three-step synthesis represents a hydrazine reduction. Using a CatCart filled with 10 % Pd/C, full conversion was achieved at 40 °C reaction temperature within several minutes residence time [47].

In a chemo-enzymatic two-step synthesis, various 3-arylated 3,4-dihydrocoumarines were synthesized. Prior to the laccase-catalyzed oxidation/Michael addition sequence, a catalytic flow hydrogenation of few 3-substituted coumarins was developed. Processing the substrate solution in THF of CH_2Cl_2 over a bed of 20 % $Pd(OH)_2/C$ at 20 °C reaction temperature and 25–40 bar hydrogen pressure, the expected 3,4-dihydrocoumarines were obtained in nearly quantitative yields [48].

The selective reduction of retronecine into platynecine was also easily achievable under flow conditions [49]. In a semisynthetic approach, monocrotaline—a natural pyrrolizidine alkaloid—was extracted from *Crotalaria spectabilis*. Subjecting it to microwave heating under basic conditions provided retronecine as the starting material for the following hydrogenation studies. RaneyNi catalysis in hydrogen atmosphere and 50 °C reaction temperature provided exclusively platynecine—a valuable pyrrolizidine alkaloid.

The reduction of numerous aromatic compounds—benzene, furan, and pyridine derivatives—could be completed by using 10 % Rh/C or 10 % Ru/C as catalyst [50]. The process was easily scaled up to prepare 10 mmol of product. A range of different conditions—temperature and hydrogen pressure—was required to cover the various substrates. In the rhodium catalyzed reduction, temperatures of between 50–100 °C and 1–50 bar hydrogen pressure were applied. Using ruthenium, the reactor temperature was set to 100 °C to achieve full conversion in most of the cases.

Another continuous flow reduction methods and instruments were also developed trough the recent years. A low-temperature diisobutylaluminium hydride (DIBAL-H) reduction of methyl butyrate to butyraldehyde was evaluated back in

2008 [51]. In comparison to batch experiments, the continuous flow concept allowed working at −20 °C instead of −78 °C as traditionally used. The same reaction selectivity was observed. Another hydride—lithium diisobutyl-*tert*-butoxyaluminum hydride (LDBBA)—was employed in a selective ester-to-aldehyde reduction in a further synthetic example [52]. Aromatic, aliphatic, heteroaromatic, and heteroaliphatic aldehydes were obtained in good to excellent yields under flow conditions and excellent selectivity. Using prepacked cartridges containing a mixture of Celite, sodium borohydride, and lithium chloride, a continuous reduction of aldehydes and ketones was realized together with several examples of reductive amination [53]. The used stoichiometric amounts of borohydride can be seen as a drawback, since after each reaction the cartridges have to be replaced. Nevertheless, the method adds some convenience in comparison to standard batch methods. A "tube-in-tube" flow device was tested for the hydrogenation of olefins under flow conditions. Here, the hydrogen gas is transported via the walls of a gas-permeable fluoropolymer tubing into the reaction stream containing the olefin [54]. A nitro-reduction of halo-substituted nitroaromatics as an important step in the synthesis of clofazimine and vismodegib was reported in another example [55]. A hydrogen gas bottle was used in the optimized flow process. Several commercially available heterogeneous catalysts were tested before achieving the required selectivity for the nitro-reduction with RaneyNi. Another instrumental setup—the EYELA CCR-1000G flow reactor— was also evaluated in the nitro-reduction reaction of a set of nitroaromatics [56]. Hydrogen was delivered to the reaction mixture from an external bottle.

Without using heterogeneous metal catalyst and hydrogen gas, the reduction of various aldehydes was achieved in flow by a transfer hydrogenation process [57]. As hydrogen donor, 2-propanol was used in combination with lithium isopropoxide. To accelerate the process, high reaction temperatures were required—temperature of 180 °C allowed substrate conversion within 30 min. A small set of aromatic and aliphatic ketones was reduced to the corresponding alcohols. In a similar approach, using partially hydrated zirconium oxide as catalyst and 2-propanol as a solvent and hydrogen donor, i.e., a flow Meerwein–Ponndorf–Verley reduction, was disclosed by the same group [58]. A plethora of substrates was subjected to the newly developed conditions. While the majority of aldehydes reacted under rather mild conditions (60 °C for 6–15 min), the tested ketones required 120–130 °C reaction temperature and 22–75 min reaction time, nevertheless delivering nearly quantitative yields in most cases. Identical transfer hydrogenation procedure was developed based on the application of an immobilized iridium Cp* catalyst [59]. Typical conditions comprised reaction temperature of 90 °C and approximately 32 min reaction time.

A potentially attractive alternative to traditional flow reduction methods was assessed recently. Nanoparticles of Fe_3O_4 were generated from $Fe(acac)_3$ in the presence of hydrazine hydrate, which were then use in situ as a catalyst for the selective nitro-reduction of various aromatic nitroarenes [60]. At 150 °C temperature, very short reaction times (2–8 min) were sufficient to obtain the corresponding anilines. The same method works also for the reduction of organic azides

[61]. Immobilization of the Fe_3O_4-nanoparticles on a solid support affords an efficient heterogeneous catalyst, which can be also conveniently packed in cartridges to ease the reaction workup by avoiding the catalyst filtration step [62]. Using hydrazine hydrate under oxidative conditions (oxygen gas) was explored by the same group for the reduction of olefins without the need of a metal catalyst in flow as well [63].

6.2 Hydrogenolysis

In a synthetic work dealing with the preparation of dihydropyrimidine C5 amides and esters, benzyl ester deprotection provided access to several C5 dihydropyrimidine acids (Scheme 6.9). Not surprisingly, the flow hydrogenation provided better yields as compared to conventional and microwave batch experiments. The improved mixing and effects of the higher surface-to-volume ratio on the catalyst surface allowed up to 30 % higher isolated yields [64, 65].

Debenzylation and Cbz-group removal were also tested at the Abbott laboratories in a method validation study [8]. Spencer et al. utilized a flow hydrogenation unit for the debenzylation of novel 1,4-benzodiazepin-2-ones that were tested for antitrypanosomal activity [66]. As the final step in the total synthesis of (±)-epigallocatechin gallate, a debromination and a total deprotection (debenzylation) were used to obtain the final product [67]. An analogous concept was adopted for the synthesis of 6-deoxy-β-D-*manno*-heptosides. Total debenzylation and azide–amine reduction were realized in a single-flow step [68]. Simultaneous multiple debenzylation was reported for the deprotection of benzyl- and benzylidene-protected carbohydrates, utilizing a continuous flow hydrogenation reactor [69, 70]. In the search for an alternative synthesis of fondaparinux sodium, continuous flow debenzylation of the corresponding polysaccharide was successful [71].

The continuous flow synthesis of di- and tripeptides has been reported. Cartridge-packed, immobilized reagents were used for the respective activation, deprotection, and coupling synthetic steps. To provide the desired tripeptides, an additional Cbz-deprotection step had to be incorporated. The dipeptide from the initial flow coupling was subjected to flow hydrogenation in an H-Cube hydrogenator. The free amine was then isolated and used for the final coupling step to obtain the tripeptide [72]. The same group used the already optimized

Scheme 6.9 Benzyl ester deprotection of a dihydropyrimidine scaffold

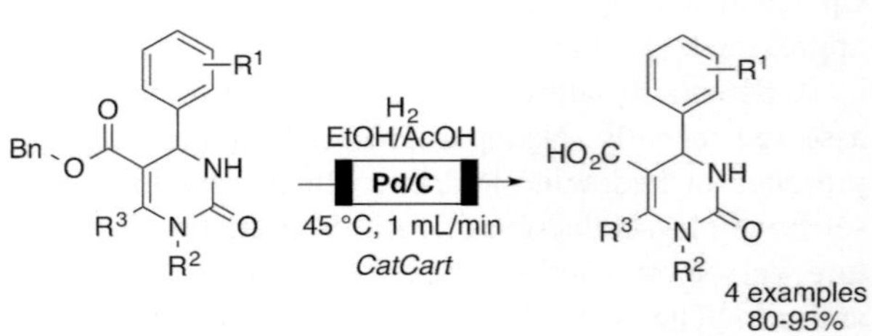

Scheme 6.10 Synthesis of (+)-chamaecypanone C involving reductive flow process

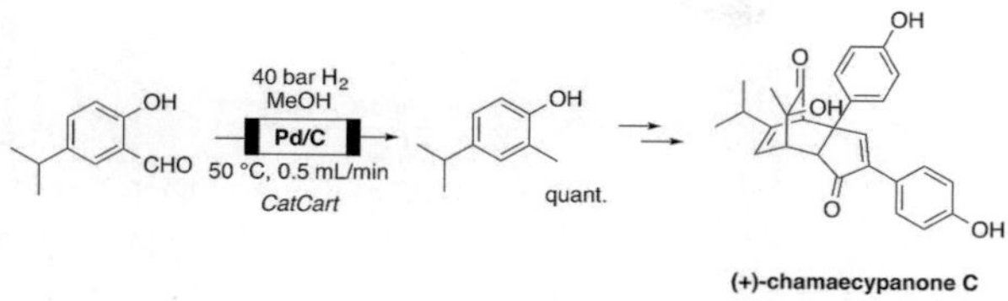

procedure for the preparation of (*S*)-pyrrolidin-2-yl-1H-tetrazole—a proline-derived organocatalyst [73]. In a flow deprotection experiment, productivity of 7 mg/min was achieved. In an extended 3.5 h run, product amount of 3 g (98 % yield) was generated. An identical batch experiment required 3 days. An optimized synthesis of (*S*)-propanolol and (S)-toliprolol involved the removal of a Cbz-protecting group from a precursor amine substrate prior to the crucial step—a high-temperature *N*-alkyl 1,2-aminoalcohol rearrangement [74]. The reductive process delivered 75 % of the desired aminoalcohol.

For the synthesis of (+)-chamaecypanone C, a natural product with microtubule inhibitory properties, reductive conditions were applied to obtain the key precursor (Scheme 6.10) [75]. The corresponding aldehyde was dissolved in methanol and passed over a bed of Pd/C at 50 °C and 40 bar hydrogen pressure to quantitatively generate the 2,4-disubstituted phenol.

A *N*-debenzylation was accomplished under reductive flow conditions in the preparation of a library of novel cyclohexanamine neuropeptide Y Y1 receptor antagonists [76]. The applied mild conditions—40 °C and 10 bar hydrogen pressure—led to quantitative deprotection. Out of a library of bicyclic β-benzyloxy amides, eight β-hydroxy amides were acquired by hydrogenolysis of the benzyl protecting group under flow conditions. All compounds were isolated in the form of single diastereomers [77]. Bridged oxa-azabicyclo [3.2.1] heterocycles were synthesized via a double-Mannich reaction of *N*,*N*-bis(methoxymethyl)-1-phenylethanamine with symmetrical and unsymmetrical ketones. One of the bicyclic products was reduced with $NaBH_4$ and the resulting alcohol reductively deprotected in the presence of Boc_2O to form a carbamate [78]. An analogous approach was utilized at Pfizer for the synthesis of the bicyclic subunit of a diazatricyclodecane agonist of the G-protein-coupled receptor 119 [79]. Methyl 2-amino-6-methoxynicotinate was synthesized using a combination of microwave and flow techniques, whereby flow *N*-debenzylation was the final step in the three-step synthesis [80]. Similarly, *N*-debenzylation was successfully achieved with the help of hydrogenation, using the H-Cube instrument. The resulting aminoalcohol was further modified [81]. Unprotected sulfonimidamides were prepared using a mild reaction protocol—at ambient temperature, the Pd/C-catalyst cleaved the benzyl carbamate group, affording the expected product in moderate yield [82]. Flow hydrogenolysis has found application also for the synthesis of high-energy density materials (HEDMs) with cage structures as promising materials in the rocket, propellant, and explosive industry [83]. Thus, the synthesis of 2,6,8,12-tetraacetyl-2,4,6,8,10,12-hexaazaisowurtzitane (TAIW) by catalytic

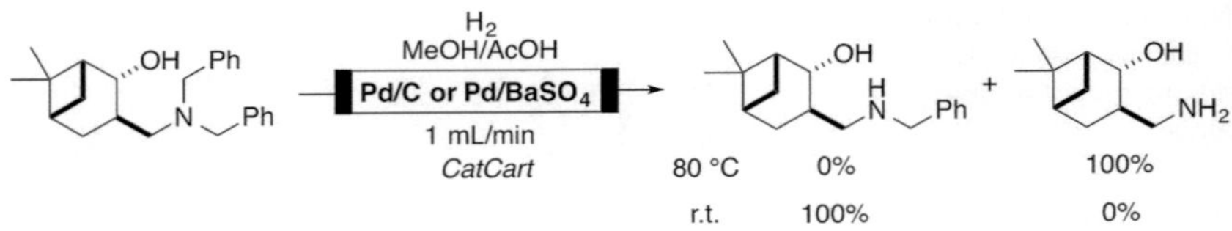

Scheme 6.11 Selective deprotection in the synthesis of 1,3-aminoalcohols

hydrogenolysis of 2,6,8,12-tetraacetyl-4,10-dibenzyl-2,4,6,8,10,12-hexaazaisowurtzitane (TADBIW), a key step in the synthesis of 2,4,6,8,10,12- hexanitro-2,4,6,8,10,12-hexaazaisowurtzitane (HNIW), has been evaluated under reductive continuous flow conditions. Several parameters (i.e., reaction temperature, flow rate, and gas pressure) have been investigated. Applying the optimized flow conditions, the isolated yield was 99 % as compared to 92 % for a batch reaction. Chiral 1,3-aminoalcohols derived from nopinone were synthesized recently [84]. A *N*-debenzylation was conducted in a flow hydrogenator at different conditions, allowing the rapid tuning of the reaction. At 80 °C working temperature and 10 % Pd/C as a catalyst, full deprotection provided the free primary 1,3-aminoalcohol. Switching to room temperature and using 5 % Pd/$BaSO_4$ as catalyst, mono-debenzylation was the predominant process and only secondary 1,3-aminoalcohol was obtained (Scheme 6.11).

A reproducible and scalable methodology for the selective debenzylation of *O*-benzyl protected cyclic hydroxamates was established [85]. Previous issues with poor selectivity and reproducibility under batch reaction conditions could be solved. Interestingly, conditions relying on the use of hydrogen gas were less effective as compared to the optimized protocol using ammonium formate (HCO_2NH_4) as hydrogen donor. When using hydrogen gas, over-reduction occurred in all cases (≥5 %). With HCO_2NH_4/SiliaCat® DPP-Pd in methanol as hydrogen/catalyst system O-debenzylation was the favored reaction pathway (≥97 %). With the optimized deprotection conditions at hand, a library of 28 *N*-hydroxypyrazin-2(1*H*)-one analogues was successfully prepared.

6.3 Reductive Amination

As an evaluation reaction for an integrated synthesis and purification system at Abbott, based on a flow synthesizer—the Accendo Conjure reactor—a reductive amination was investigated along with other reactions [86]. The Reissert indole synthesis provided series of substituted ethyl indole-2-carboxylates and aza-indole analogues under reductive continuous flow conditions (Scheme 6.12) [87].

As starting materials, *o*-nitrophenylpyruvates were used. Depending on the structure and substitution of the substrates, Pd/C, Ru/C, and RaneyNi were utilized as catalysts. Five different sets of conditions were developed to cover the whole

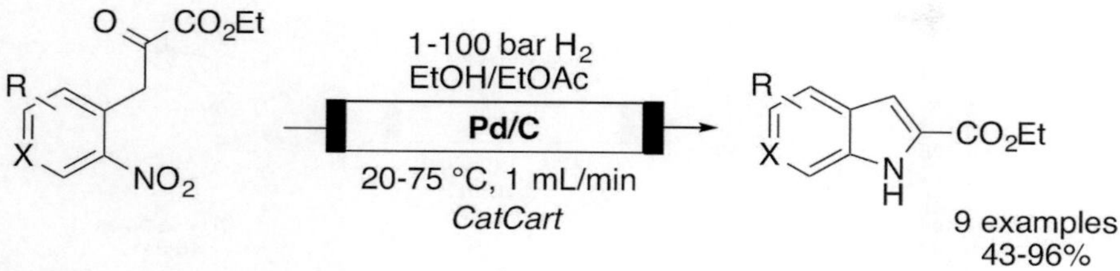

Scheme 6.12 Reissert indole synthesis under reductive continuous flow conditions

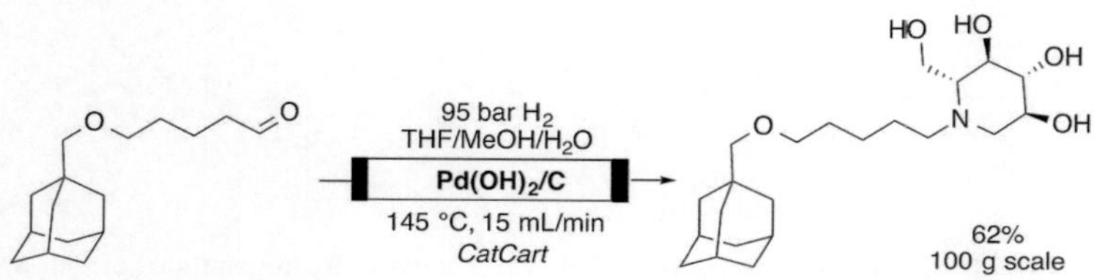

Scheme 6.13 Continuous flow synthesis of a potent glucosylceramide synthase inhibitor

range of substrates—temperatures of 20–75 °C, 1–100 bar hydrogen pressure, and 1–3 mL/min flow rate.

A comparative study demonstrating the abilities of flow processing was disclosed, focused on batch and flow reductive aminations [88]. For the batch experiments, combinations of HCO_2NH_4 and Pd/C or zinc dust were selected. For the flow experiments, the reductive system of HCO_2NH_4-Pd/C was used. Surprisingly, no obvious advantage from the use of flow equipment could be found—the outcome from the batch and flow experiments were virtually the same. Nevertheless, the apparent convenience of using a packed-bed heterogeneous catalyst remains, since the filtration step after the reaction can be omitted.

An economic, scalable process for the production of glucosylceramide synthase (GCS) inhibitor has been devised [89]. The synthetic strategy was based on the assembly of 5-adamantylmethoxy-1-pentanal and 1-deoxynojirimycin via reductive amination (Scheme 6.13), which turned out to be surprisingly difficult. The water solubility of the product hinders the reductive amination using borohydrides or transfer hydrogenation conditions, and for these reasons, catalytic hydrogenation under flow conditions was considered. The optimized conditions—150 °C, 15 mL/min flow rate, and 100 bar hydrogen pressure over $Pd(OH)_2/C$— allowed consistent production of the inhibitor on a 100 g scale.

In an industrial project, the 200–300 g flow synthesis of benzylpiperazine was designed [90]. The synthetic strategy was based on an unprotected piperazine in a reductive amination with the corresponding aldehyde where typically multiple side products are to be expected. Nevertheless, after carefully adjusting the reaction parameters—solvent, catalyst, temperature, hydrogen pressure, flow rate, and substrate concentration—selective conditions were found. Using $Pd(OH)_2/C$ as catalyst at 70 °C reaction temperature, and hydrogen atmosphere, 4 equivalents of

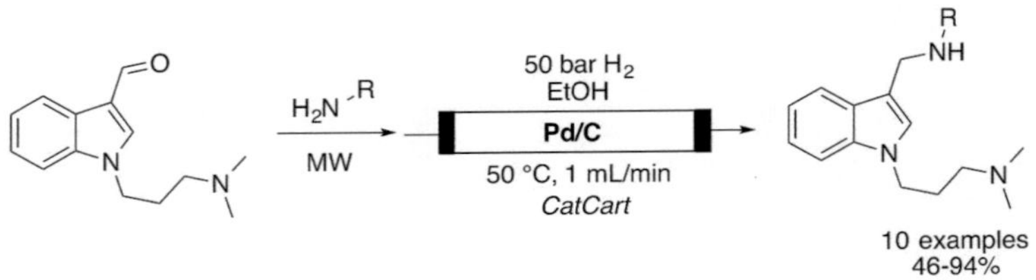

Scheme 6.14 A combined microwave-flow approach towards indole-based dynamin GTPase inhibitors

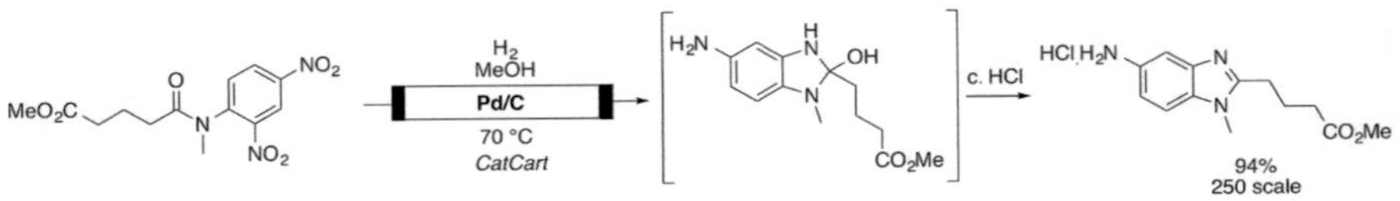

Scheme 6.15 Benzimidazole ring formation via catalytic hydrogenation of an aromatic-compound

unprotected piperazine and a 0.5 M solution of the aldehyde in methanol were processed to provide the desired amination product in 98 % selectivity and full conversion. Only minor amounts of the double reductive amination products were detected.

Focused library development of indole-based dynamin GTPase inhibitors was reported by McCluskey et al. [91]. Additional molecules were generated by a simple reductive amination of indole carboxaldehyde with a series of amines in an H-cube flow hydrogenator. The starting imine was generated by microwave heating of a suspension of alkylamine, substituted indole-3-carboxaldehyde, and $MgSO_4$ in toluene for 10 min. After workup, the obtained solution was further diluted and subjected to flow reduction (50 °C temperature, 50 bar hydrogen pressure, 1 mL/min flow rate) (Scheme 6.14). The sequential microwave/flow synthesis provided a number of novel compounds with highly promising receptor inhibition activity in the μM-range.

A new synthetic route for the synthesis of bendamustine hydrochloride based on a continuous flow reductive amination was established [92]. The key benzimidazole intermediate was generated by a simultaneous intramolecular double nitro-reduction/cyclization (Scheme 6.15). The dehydration was subsequently conducted in the presence of conc. HCl. The optimized process was demonstrated on a 250 g scale.

Besides reduction of aldehydes and ketones, continuous flow reductive amination using a packed column with specially formulated solid $NaBH_4$ was demonstrated by the group of Seeberger [93]. The concept circumvents the typical clogging issues encountered while working with solids under flow conditions. Although successful, the concept suffers from the fact that the used borohydride leaches in stoichiometric amounts, and after a certain period of time, the used column has to be replaced by a fresh one. This means an interruption of the flow process. From a technical point of view, a solution for the problem can be the use of at least two cartridges in parallel, allowing the on-line change or recharge without the mentioned stoppage.

The continuous synthesis of 1,4-benzodiazepin-5-ones, a privileged scaffold for drug discovery, has been achieved ([94], see also [36]). Using 2-nitro benzamides as starting materials in a continuous reductive amination, the desired heterocyclic scaffolds were conveniently obtained in high yields.

6.4 Oxidations

Oxidation reactions play a crucial role in organic synthesis as well. Various processes have been reported, exploiting the special methodological features of flow processing to establish a simplified and safer synthetic procedure to generate aldehydes, ketones, carboxylic acids, peroxides, etc.

Back in 2008, a spinning tube-in-tube reactor developed and marketed by Kreido Biofuels was used for TEMPO-catalyzed oxidations of alcohols by hypochlorite. Continuous production of various aldehydes or ketones was possible at 4000–6000 RPMs and residence times of 0.7–3 min at 0 °C reaction temperature. Secondary alcohols reacted somewhat slower, as known for TEMPO-catalyzed oxidations [95]. A general procedure for TEMPO-mediated electrooxidations of primary and secondary alcohols was also established in flow environment [96]. The mild conditions at ambient temperature provide effective access to various aldehydes and ketones without the need for additional electrolyte and using only a buffered aqueous *tert*-butanol as the reaction milieu. The 15 substrates were converted into aldehydes and ketones in 21–98 % isolated yields. A TEMPO/iron oxide cooperative catalysis was evaluated for the selective oxidation of benzyl alcohol into benzaldehyde [97].

However, the most attractive oxidation processes are based on the use of molecular oxygen as an oxidant, leaving only water as a side product. Hii and coworkers reported such a process for the oxidation of alcohols using an X-Cube system [98]. Besides primary alcohols, all the tested substrates were quantitatively converted into aldehydes or ketones. Using a packed-bed with 5 % Ru/Al_2O_3 as a catalyst at 90 °C and 5 bar of oxygen gas pressure at 1 mL/min flow rate, the oxidation process was completed in only 44 s residence time in the cartridge. In several cases, oxygen could be replaced by compressed air without negatively affecting the oxidation performance. Safe operation and simplified workup characterize the overall process. The aerobic flow oxidation of alcohols was also elaborated in a recent work by Uozomi and coworkers [99]. In an aqueous reaction system with packed platinum nanoparticles dispersed on an amphiphilic polystyrene–poly(ethylene glycol) resign as catalyst, efficient oxidation of various primary and secondary aliphatic, aromatic and heteroaromatic alcohols into the corresponding carboxylic acids and ketones was achieved.

Another very often used and cost-effective oxidation reagent is $KMnO_4$. An extended study on potassium permanganate oxidations of alcohols and aldehydes to

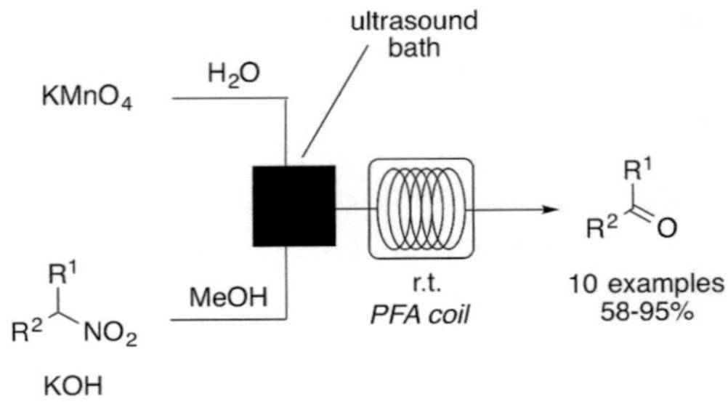

Scheme 6.16 The Nef oxidation towards aldehydes/ketones

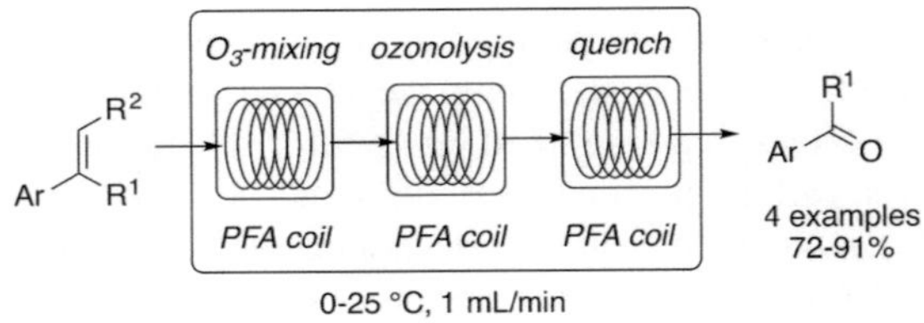

Scheme 6.17 Ozonolysis in the flow generation of aromatic ketones

carboxylic acids and nitroalkanes to the corresponding carbonyls and carboxylic acids (Nef oxidation) was recently published [100]. A problematic issue was the processing of the generated MnO_2 that caused blockages of the flow equipment. To overcome this issue, an ultrasound bath had to be used (Scheme 6.16). A large number of substrates was subjected to the optimized conditions to obtain a library of diversely substituted ketones and carboxylic acids.

A different interesting oxidizing reagent is ozone. However, the practical use of ozone is rather restricted due to major safety concerns since the formed intermediates are unstable and potentially explosive. Thus, an efficient temperature control of the overall process is of great importance. A continuous flow laboratory device was launched on the market several years ago (O-Cube^{TM}, Thalesnano), able to address these issues. The instrument implements both the ozonolysis and the quenching steps in a flow environment. Diversely substituted alkenes (Scheme 6.17) as well as alkynes, amines, and thiols were easily oxidized under corresponding conditions [101].

Interestingly, most of the transformations worked well under room temperature conditions. In the case of thioanisole, precise temperature control allowed the selective production of either the respective sulfoxide or sulfone. The selective oxidation of sulfides into sulfoxides or sulfones is a challenging and technologically demanding process and has been very recently reevaluated using hydrogen peroxide in combination with a peroxometalate-based polymer immobilized ionic liquid phase catalyst—$[PO_4\{WO(O_2)_2\}_4]$@PIILP [102]. The pursued selectivity towards sulfoxides or sulfones was achieved by a solvent switch and residence time tuning. Using methanol as a solvent and short residence time favored the formation of the

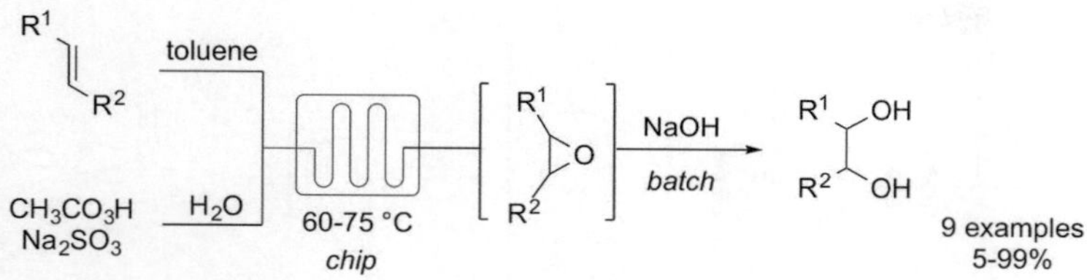

Scheme 6.18 Prilezhaev epoxidation/diol synthesis under continuous flow conditions

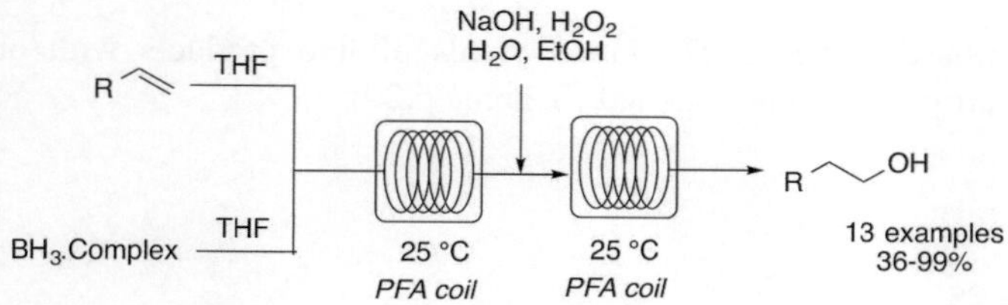

Scheme 6.19 Continuous flow hydroboration–oxidation sequence

desired sulfoxides, while the use of acetonitrile as the solvent and extended reaction time resulted in the almost exclusive formation of the corresponding sulfone.

A polystyrene-based monolithic version of N-(*tert*-butyl)phenylsulfinimidoyl chloride was utilized as an immobilized oxidizing reagent to transform a broad range of alcohols into the corresponding aldehydes or ketones [103]. The reagent is applicable in both stoichiometric and catalytic amounts and can be easily regenerated. The same research group evaluated partially hydrated zirconium oxide as a packed-bed catalyst in the presence of oxidants such as acetone, cyclohexanone, or neopentanal for the oxidation of benzylic secondary alcohols (Oppenauer oxidation) [104]. The reaction conditions (40–100 °C, 12–30 min reaction time) worked well for both electron-rich and electron-deficient substrates.

Epoxidation of both terminal and internal olefins via the Prilezhaev reaction utilizing peroxyacetic acid was also evaluated and coupled with a basic quench in batch to hydrolyze the formed epoxide affording various diols [105]. An attempt to perform both steps under flow conditions failed due to the repeated clogging of the reactor. In most cases, reaction temperature of 60–75 °C was sufficient to obtain the desired products within 5–10 min reaction times (Scheme 6.18).

A continuous flow protocol has been optimized for the hydroboration–oxidation of olefins [106]. Using a THF-BH_3 complex for the hydroboration step and a combination of hydrogen peroxide–ethanol–water in the presence of a base for the oxidation step delivered the best conversions during the optimization process (Scheme 6.19). The reaction was performed at ambient temperature. The final two-stage flow procedure incorporated also an in-line liquid–liquid phase separation.

Combining ozonolysis with amide bond formation by aminolysis of oxazolidinone, the synthesis of some adamantane analogues as potential $P2X_7$-

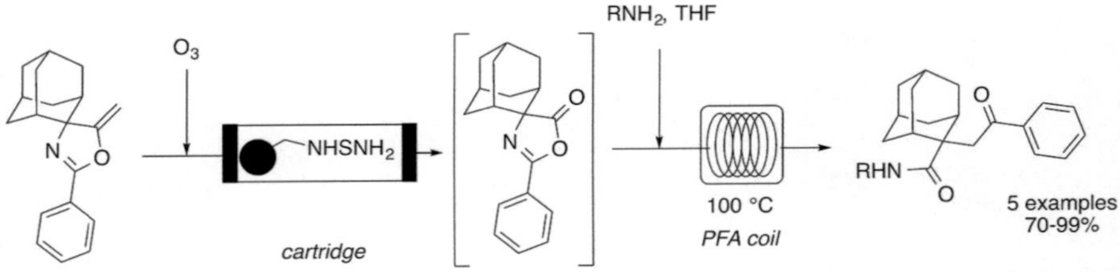

Scheme 6.20 Ozonolysis/aminolysis flow sequence for the synthesis of $P2X_7$-inhibitors

inhibitors was achieved [107]. Good yields of five products with outstanding analgesics properties were obtained (Scheme 6.20).

References

1. S. Saaby, K.R. Knudsen, M. Ladlow, S.V. Ley, Chem. Commun. **5**, 2909 (2005)
2. R.V. Jones, L. Godorhazy, N. Varga, D. Szalay, L. Urge, F. Darvas, J. Comb. Chem. **8**, 110 (2006)
3. I. Kovacs, R.V. Jones, K. Niesz, C. Csajagi, B. Borcsek, F. Darvas, L. Urge, J. Lab. Autom. **12**, 284 (2007)
4. M.C. Bryan, D. Wernick, C.D. Hein, J.V. Petersen, J.W. Eschelbach, E.M. Doherty, Beilstein J. Org. Chem. **7**, 1141 (2011)
5. M. Irfan, T.N. Glasnov, C.O. Kappe, ChemSusChem **4**, 300 (2011)
6. J.W. Eschelbach, D. Wernick, M.C. Bryan, E.M. Doherty, Aust. J. Chem. **66**, 165 (2013)
7. T. Hattori, A. Tsubone, Y. Sawama, Y. Monguchi, H. Sajiki, Tetrahedron **70**, 4790 (2014)
8. B. Clapham, N.S. Wilson, M.J. Michmerhuizen, D.P. Blanchard, D.M. Dingle, T.A. Nemcek, J.Y. Pan, D.R. Sauer, J. Comb. Chem. **10**, 88 (2008)
9. H.H. Horvath, G. Papp, C. Csajagi, F. Joo, Catal. Commun. **8**, 442 (2007)
10. S. Balogh, G. Farkas, J. Madarasz, A. Szöllösy, J. Kovacs, F. Darvas, L. Ürge, J. Bakos, Green Chem. **14**, 1146 (2012)
11. G. Vile, N. Almora-Barrios, S. Mitchell, N. Lopez, J. Perez-Ramirez, Chem. Eur. J. **20**, 5926 (2014)
12. G. Vile, N. Almora-Barrios, N. Lopez, J. Perez-Ramirez, ACS Catal. **5**, 3767 (2015)
13. T. Osako, K. Torii, A. Tazawa, Y. Uozumi, RSC Adv. **5**, 45760 (2015)
14. G. Szöllosi, S. Cserenyi, F. Fülöp, M. Bartok, J. Catal. **260**, 245 (2008)
15. G. Szöllosi, S. Cserenyi, K. Balazsik, F. Fülöp, M. Bartok, J. Mol. Catal. A **305**, 155 (2009)
16. G. Szöllosi, S. Cserenyi, K. Szöri, F. Fülöp, M. Bartok, Catal. Commun. **12**, 14 (2010)
17. G. Szöllosi, S. Cserenyi, M. Bartok, Catal. Lett. **134**, 264 (2010)
18. G. Szöllosi, Z. Makra, F. Fülöp, M. Bartok, Catal. Lett. **141**, 1616 (2011)
19. G. Szöllosi, Z. Makra, M. Fekete, F. Fülöp, M. Bartok, Catal. Lett. **142**, 889 (2012)
20. G. Arnott, H. Brice, J. Clayden, E. Blaney, Org. Lett. **10**, 3089 (2008)
21. M. Irfan, E. Petricci, T.N. Glasnov, M. Taddei, C.O. Kappe, Eur. J. Org. Chem. **2009**, 1327 (2009)
22. I.M. Mandity, T.A. Martinek, F. Darvas, F. Fülöp, Tetrahedron Lett. **50**, 4372 (2009)
23. S.B. Ötvös, I.M. Mandity, F. Fülöp, Mol. Divers. **15**, 605 (2011)
24. S. Chandrasekhar, B.V.D. Vijaykumar, B.M. Chandra, C.R. Reddy, P. Naresh, Tetrahedron Lett. **52**, 3865 (2011)

25. C.-T. Hsieh, S.B. Ötvös, Y.-C. Wu, I.M. Mandity, F.-R. Chang, F. Fülöp, ChemPlusChem **5**, 859 (2015)
26. C.J. O'Brien, J.L. Tellez, Z.S. Nixon, L.J. Kang, A.L. Carter, S.R. Kunkel, K.C. Przeworski, G.A. Chass, Angew. Chem. Int. Ed. **48**, 6836 (2009)
27. M.D.H. Bhuiyan, A.B. Mahon, P. Jensen, J.K. Clegg, A.C. Try. Eur. J. Org. Chem. 687 (2009)
28. T.N. Glasnov, C.O. Kappe, Adv. Synth. Catal. **352**, 3089 (2010)
29. H. Huang, D.A. Hutta, J.M. Rinker, H. Hu, W.H. Parsons, C. Schubert, R.L. DesJarlais, C.S. Crysler, M.A. Chaikin, R.R. Donatelli, Y. Chen, D. Cheng, Z. Zhou, E. Yurkow, C.L. Manthey, M.R. Player, J. Med. Chem. **52**, 1081 (2009)
30. H.R. Lawrence, A. Kazi, Y. Luo, R. Kendig, Y. Ge, S. Jain, K. Daniel, D. Santiago, W.C. Guida, S.M. Sebti, Bioorg. Med. Chem. **18**, 5576 (2010)
31. D. Obermayer, T.N. Glasnov, C.O. Kappe, J. Org. Chem. **76**, 6657 (2011)
32. G.G. de la Cruz, K. Groschner, C.O. Kappe, T.N. Glasnov, Tetrahedron Lett. **53**, 3731 (2012)
33. S.H. Liang, T.L. Collier, B.H. Rotstein, R. Lewis, M. Steck, N. Vasdev, Chem. Commun. **49**, 8755 (2013)
34. B. Desai, K. Dixon, E. Farrant, Q. Feng, K.R. Gibson, W.P. van Hoorn, J. Mills, T. Morgan, D.M. Parry, M.K. Ramjee, C.N. Selway, G.J. Tarver, G. Whitlock, A.G. Wright, J. Med. Chem. **56**, 3033 (2013)
35. Z. Qian, I.R. Baxendale, S.V. Ley, Synlett **4**, 505 (2010)
36. M. Baumann, I.R. Baxendale, C.H. Hornung, S.V. Ley, M.V. Rojo, K.A. Roper, Molecules **19**, 9736 (2014)
37. M. Viviano, C. Milite, D. Rescigno, S. Castellano, G. Sbardella, RSC Adv. **5**, 1268 (2015)
38. B. Gutmann, J.-P. Roduit, D. Roberge, C.O. Kappe, Chem. Eur. J. **17**, 13146 (2011)
39. M.C. Bryan, C.D. Hein, H. Gao, X. Xia, H. Eastwood, B.A. Bruenner, S.W. Louie, E.M. Doherty, ACS Comb. Sci. **15**, 503 (2013)
40. M. Tarleton, K.A. Young, E. Unicomb, S.N. McCluskey, M.J. Robertson, C.P. Gordon, A. McCluskey, Lett. Drug Des. Discov. **8**, 568 (2011)
41. M. Tarleton, A. McCluskey, Tetrahedron Lett. **52**, 1583 (2011)
42. J. Spencer, H. Patel, J. Amin, S.K. Callear, S.J. Coles, J.J. Deadman, C. Furman, R. Mansouri, P. Chavatte, R. Millet, Tetrahedron Lett. **53**, 1656 (2012)
43. L. Hizartzidis, M. Tarleton, C.P. Gordon, A. McCluskey, RSC Adv. **4**, 9709 (2014)
44. M. Viviano, T.N. Glasnov, B. Reichart, G. Tekautz, C.O. Kappe, Org. Process Res. Dev. **15**, 585 (2011)
45. P.M. Mirzayans, R.H. Pouwer, C.M. Williams, P.V. Bernhardt, Eur. J. Org. Chem. **2012**, 1633 (2012)
46. K. Tchabanenko, C. Sloan, Y.-M. Bunetel, P. Mullen, Org. Biomol. Chem. **10**, 4215 (2012)
47. L. Dalla-Vechia, B. Reichart, T. Glasnov, L.S.M. Miranda, C.O. Kappe, R.O.M.A. de Souza, Org. Biomol. Chem. **11**, 6806 (2013)
48. S. Suljic, J. Pietruszka, Adv. Synth. Catal. **356**, 1007 (2014)
49. S.T. Martinez, A.C. Pinto, T. Glasnov, C.O. Kappe, Tetrahedron Lett. **55**, 4181 (2014)
50. T. Hattori, T. Ida, A. Tsubone, Y. Sawama, Y. Monguchi, H. Sajiki, Eur. J. Org. Chem. **2015**, 2492 (2015)
51. L. Ducry, D.M. Roberge, Org. Process Res. Dev. **12**, 163 (2008)
52. J. de M. Munoz, J. Alcazar, A. de la Hoz, A. Diaz-Ortiz, Eur. J. Org. Chem. **2012**, 260 (2012)
53. K. Gilmore, S. Vukelic, D.T. McQuade, B. Koksch, P.H. Seeberger, Org. Process Res. Dev. **18**, 1771 (2014)
54. M.A. Mercadante, C.B. Kelly, C.X. Lee, N.E. Leadbeater, Org. Process Res. Dev. **16**, 1064 (2012)
55. P. Loos, H. Alex, J. Hassfeld, K. Lovis, J. Platzek, N.Steinfeldt, S. Hübner, Org. Process Res. Dev. (2015). doi: 10.1021/acs.oprd.5b00170
56. M. Ueno, Y. Morii, K. Uramoto, H. Oyamada, Y. Mori, S. Kobayashi, J. Flow Chem. **4**, 160 (2014)

57. J. Sedelmeier, S.V. Ley, I.R. Baxendale, Green Chem. **11**, 683 (2009)
58. C. Battilocchio, J.M. Hawkins, S.V. Ley, Org. Lett. **15**, 2278 (2013)
59. M.V. Rojo, L. Guetzoyan, I.R. Baxendale, Org. Biomol. Chem. **13**, 1768 (2015)
60. D. Cantillo, M. Baghbanzadeh, C.O. Kappe, Angew. Chem. Int. Ed. **51**, 10190 (2012)
61. D. Cantillo, M.M. Moghaddam, C.O. Kappe, J. Org. Chem. **78**, 4530 (2013)
62. M.M. Moghaddam, B. Pieber, T. Glasnov, C.O. Kappe, ChemSusChem **7**, 3122 (2014)
63. B. Pieber, S.T. Martinez, D. Cantillo, C.O. Kappe, Angew. Chem. Int. Ed. **52**, 10241 (2013)
64. B. Desai, C.O. Kappe, J. Comb. Chem. **7**, 641 (2005)
65. B. Desai, D. Dallinger, C.O. Kappe, Tetrahedron **62**, 4651 (2006)
66. J. Spencer, R.P. Rathnam, A.L. Harvey, C.J. Clements, R.L. Clark, M.P. Barrett, P.E. Wong, L. Male, S.J. Coles, S.P. Mackay, Bioorg. Med. Chem. **19**, 1802 (2011)
67. H. Tanaka, A. Chino, T. Takahashi, Tetrahedron Lett. **53**, 2493 (2012)
68. M.T. Kenfack, Y. Bleeriot, C. Gauthier, J. Org. Chem. **79**, 4615 (2014)
69. F.S. Ekholm, I.M. Mandity, F. Fülöp, R. Leino, Tetrahedron Lett. **52**, 1839 (2011)
70. R. Leyden, T. Velasco-Torrijos, S. Andre, S. Gouin, H.-J. Gabius, P.V. Murphy, J. Org. Chem. **74**, 9010 (2009)
71. A. Manikowski, A. Koziol, E. Czajkowska-Wojciechowska, Carbohydr. Res. **361**, 155 (2012)
72. I.R. Baxendale, S.V. Ley, C.D. Smith, G.K. Tranmer, Chem. Commun. **2006**, 4835 (2006)
73. V. Franckevicius, K.R. Knudsen, M. Ladlow, D.A. Longbottom, S.V. Ley, Synlett **17**, 889 (2006)
74. B. Duthion, T.-X. Metro, D.G. Pardo, J. Cossy, Tetrahedron **65**, 6696 (2009)
75. S. Dong, E. Hamel, R. Bai, D.G. Covell, J.A. Beutler, J.A. Porco Jr., Angew. Chem. Int. Ed. **48**, 1494 (2009)
76. K. Cho, M. Ando, K. Kobayashi, H. Miyazoe, T. Tsujino, S. Ito, T. Suzuki, T. Tanaka, S. Tokita, N. Sato, Bioorg. Med. Chem. Lett. **19**, 4781 (2009)
77. L. Zhang, Q. Xiao, C. Ma, X.-Q. Xie, P.E. Floreancig, J. Comb. Chem. **11**, 640 (2009)
78. E. Darout, A. Basak, Tetrahedron Lett. **51**, 2998 (2010)
79. E. Darout, R.P. Robinson, K.F. McClure, M. Corbett, B. Li, A. Shavnya, M.P. Andrews, C.S. Jones, Q. Li, M.L. Minich, V. Mascitti, C.R.W. Guimaraes, M.J. Munchhof, K.B. Bahnck, C. Cai, D.A. Price, S. Liras, P.D. Bonin, P. Cornelius, R. Wang, V. Bagdasarian, C.P. Sobota, S. Hornby, V.M. Masterson, R.M. Joseph, A.S. Kalgutkar, Y. Chen, J. Med. Chem. **56**, 301 (2013)
80. G. Jeges, T. Meszaros, T. Szommer, J. Kovacs, T. Nagy, D. Tymoshenko, N. Fotouhi, P. Gillespie, A. Kowalczyk, R.A. Goodnow Jr., Synlett 203 (2011)
81. N.M. Horwath, J. Ricci, Tetrahedron **67**, 9588 (2011)
82. M.F. Maldonado, F. Sehgelmeble, F. Bjarnemark, M. Svensson, J. Ahman, P.I. Arvidsson, Tetrahedron **68**, 7456 (2012)
83. K. Dong, C.H. Sun, J.W. Song, G.X. Wei, S.P. Pang, Org. Process Res. Dev. **18**, 1321 (2014)
84. Z. Szakonyi, T. Gonda, S.B. Ötvös, F. Fülöp, Tetrahedron Asymmetry **25**, 1138 (2014)
85. A.H. Mai, W.M. De Borggraeve, J. Flow Chem. **5**, 6 (2015)
86. J.E. Hochlowski, P.A. Searle, N.P. Tu, J.Y. Pan, S.G. Spanton, S.W. Djuric, J. Flow Chem. **2**, 56 (2011)
87. E. Colombo, P. Ratel, L. Mounier, F. Guillier, J. Flow Chem. **2**, 68 (2011)
88. P. Falus, Z. Boros, G. Hornyanszky, J. Nagy, F. Darvas, L. Ürge, L. Poppe, Tetrahedron Lett. **52**, 1310 (2011)
89. C.G.F. Cooper, E.R. Lee, R.A. Silva, A.J. Bourque, S. Clark, S. Katti, V. Nivorozhkin, Org. Process Res. Dev. **16**, 1090 (2012)
90. J. Liu, E. Fitzgerald, N.S. Mani, Synthesis **44**, 2469 (2012)
91. C.P. Gordon, B. Venn-Brown, M.J. Robertson, K.A. Young, N. Chau, A. Mariana, A. Whiting, M. Chircop, P.J. Robinson, A. McCluskey, J. Med. Chem. **56**, 46 (2013)
92. J. Chen, K. Przyuski, R. Roemmele, R.P. Bakale, Org. Process Res. Dev. **18**, 1427 (2014)
93. K. Gilmore, S. Vukelicc, D.T. McQuade, B. Koksch, P.H. Seeberger, Org. Process Res. Dev. **18**, 1771 (2014)

94. M. Viviano, C. Millite, D. Rescigno, S. Castellano, G. Sbardella, RSC Adv. **5**, 1268 (2015)
95. P.D. Hampton, M.D. Whealon, L.M. Roberts, A.A. Yaeger, R. Boydson, Org. Process Res. Dev. **12**, 946 (2008)
96. J.T. Hill-Cousins, J. Kuleshova, R.A. Green, P.R. Birkin, D. Pletcher, T.J. Underwood, S.G. Leach, R.C.D. Brown, ChemSusChem **5**, 326 (2012)
97. D. Obermayer, A.M. Balu, A.A. Romero, W. Goessler, R. Luque, C.O. Kappe, Green Chem. **15**, 1530 (2013)
98. N. Zotova, K. Hellgardt, G.H. Kelsall, A.S. Jessiman, K.K. Hii, Green Chem. **12**, 2157 (2010)
99. T. Osako, K. Torii, Y. Uozumi, RSC Adv. **5**, 2647 (2015)
100. J. Sedelmeier, S.V. Ley, I.R. Baxendale, M. Baumann, Org. Lett. **12**, 3618 (2010)
101. M. Irfan, T.N. Glasnov, C.O. Kappe, Org. Lett. **13**, 984 (2011)
102. S. Doherty, J.G. Knight, M.A. Carroll, J.R. Ellison, S.J. Hobson, S. Stevens, C. Hardacre, P. Goodrich, Green Chem. **17**, 1559 (2015)
103. H. Lange, M.J. Capener, A.X. Jones, C.J. Smith, N. Nikbin, I.R. Baxendale, S.V. Ley, Synlett 869 (2011)
104. R. Chorghade, C. Battilocchio, J.M. Hawkins, S.V. Ley, Org. Lett. **15**, 5698 (2013)
105. B.A.M.W. van den Broek, R. Becker, F. Kössl, M.M.E. Delville, P.J. Nieuwland, K. Koch, F.P.J.T. Rutjes, ChemSusChem **5**, 289 (2012)
106. J.A. Souto, R.A. Stockman, S.V. Ley, Org. Biomol. Chem. **13**, 3871 (2015)
107. C. Battilocchio, L. Guetzoyan, C. Cervetto, L.D.C. Mannelli, D. Frattaroli, I.R. Baxendale, G. Maura, A. Rossi, L. Sautebin, M. Biava, C. Ghelardini, M. Marcoli, S.V. Ley, ACS Med. Chem. Lett. **4**, 704 (2013)

Chapter 7
Organic Synthesis in Dedicated Continuous Flow Systems

Heterocyclic and Multistep Syntheses

7.1 Heterocyclic Syntheses

Schwalbe et al. developed a multistep microreactor approach to prepare a library of fluoroquinolone antibiotics such as Ciprofloxacin [1]. Five sequential microreactor transformations delivered the desired products. Several Ciprofloxacin analogues were synthesized in good yields and purities. Ciprofloxacin itself could be synthesized in 57 % overall yield and a purity exceeding 90 % after five steps.

Using a commercial flow system, Acke and Stevens prepared a series of 3,4-diamino-1*H*-isochromen-1-ones via a modified Strecker reaction, requiring the use of highly toxic HCN [2]. The reagent solutions—2-formylbenzoic acid in acetic acid and aromatic amines with potassium cyanide—were introduced into the reactor, leading to in situ formation of HCN and an imine within the microreactor. After a total reaction time of 40 min, the corresponding isochromenones were obtained in 49–75 % yield (Scheme 7.1).

Various thiazoles/imidazoles were prepared using several isothiocyanates and ethyl isocyanoacetate in a modular flow reactor [3]. Only small amounts of the desired thiazoles were obtained. By flowing an α-bromoketone through the PS-BEMP column, the residual material was eluted as the regioisomeric imidazole, resulting in combined yields of 79–99 %. Upon reacting carbon disulfide with alkyl isothiocyanates, only the corresponding thiazoles were obtained (Scheme 7.2).

Stevens and coworkers recently presented the continuous preparation of 1*H*-isochromeno[3,4-*d*]imidazol-5-ones from 3-amino-4-(arylamino)-1*H*-isochromen-1-ones recently [4]. The ring closure was achieved by reacting the starting heterocycles, orthoester, and catalytic amounts of *p*-TsOH at room temperature. With the optimized conditions at hand, nine different 1*H*-isochromeno[3,4-*d*]imidazol-5-ones were prepared in moderate to good yields (Scheme 7.3).

Tricyclic spiropiperidines as building blocks for the histrionicotoxin family of alkaloids were prepared in a flow domino cyclization reaction of a bis-unsaturated

T. Glasnov, *Continuous-Flow Chemistry in the Research Laboratory*,
DOI 10.1007/978-3-319-32196-7_7

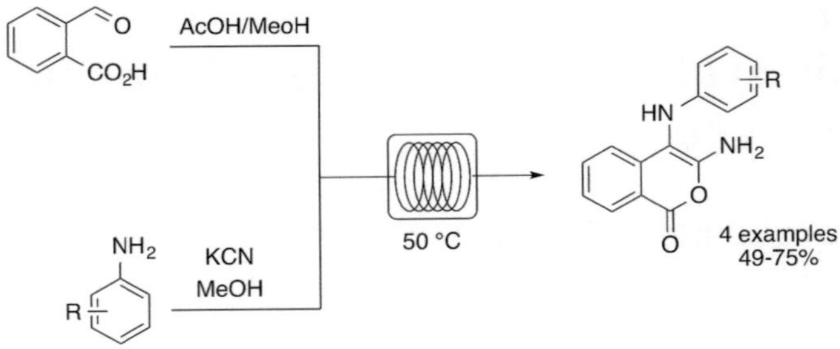

Scheme 7.1 Synthesis of 3,4-diamino-1*H*-isochromen-1-ones via a modified Strecker reaction under flow conditions

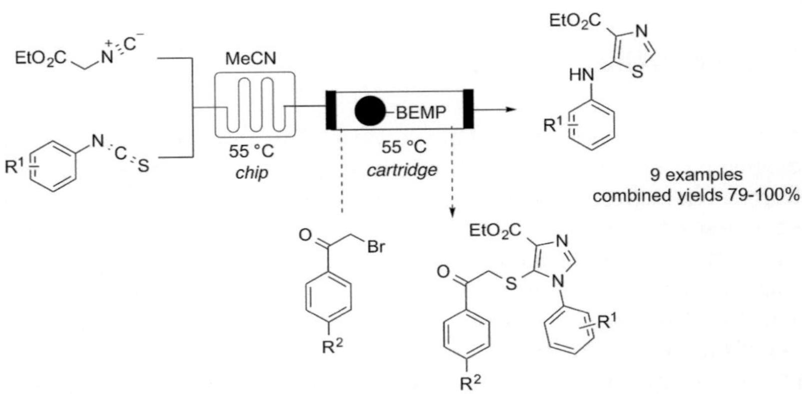

Scheme 7.2 Thiazoles/imidazoles synthesis under continuous flow conditions

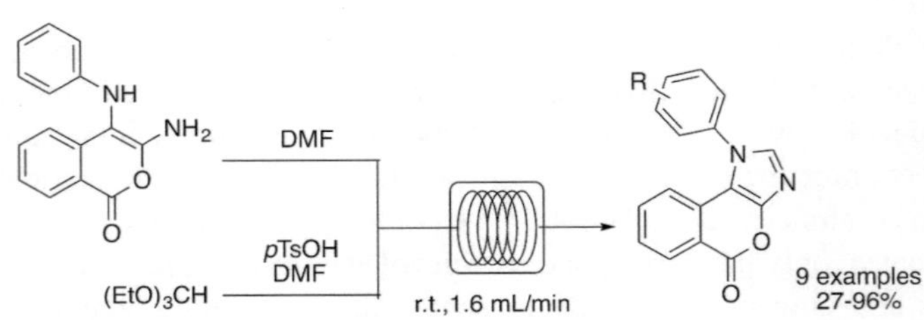

Scheme 7.3 Synthesis of different 1*H*-isochromeno[3,4-*d*]imidazol-5-ones under flow conditions

ketone with hydroxylamine and proceeded with good yields and high stereoselectivity (Scheme 7.4) [5].

A library of drug-like trisubstituted pyrrolidines was prepared using microreactor technology [6]. A sequential cyclization–reduction–amidation process was developed for the preparation of the desired pyrrolidines. The classical Hantzsch 1,4-dihydropyridine synthesis was examined in a comparative study between a microwave batch and a continuous flow equipment [7]. In a similar

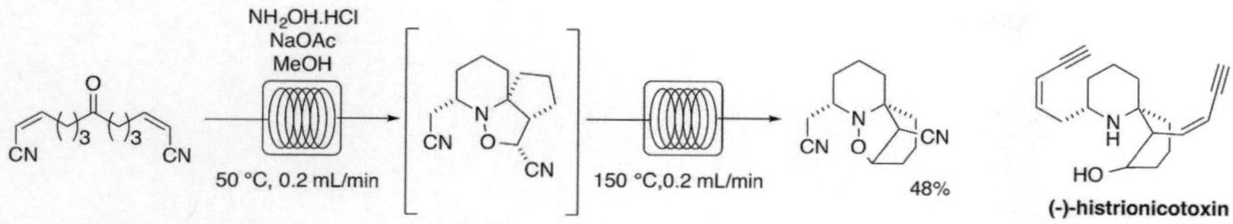

Scheme 7.4 Flow synthesis of tricyclic spiropiperidines as building blocks for histrionicotoxin alkaloid analogues

heterocyclic process—the Biginelli dihydropyrimidine condensation—a sequential β-ketothiazole formation/multicomponent reaction was achieved using a flow reactor [8]. Sequential Hantzsch thiazole synthesis, deketalization, and Biginelli reaction provided access to highly functionalized, pharmacologically useful 5-(thiazol-2-yl)-3,4-dihydropyrimidin-2(1*H*)-ones. The complex products were rapidly generated in relatively good yields—39–46 % (Scheme 7.5).

An efficient, safe, and scalable continuous flow thermolysis of azidoacrylates to obtain various indoles has been recently developed [9]. As the final part of the study, a precursor to the DAAO inhibitor 4*H*-furo[3,2-*b*]pyrrole-5-carboxylic acid was prepared. In a scale-up experiment, 8.5 g of the corresponding ester were obtained in a flow experiment at 180 °C and only 12 s residence time (Scheme 7.6). Later on, the same research group expanded the synthesis towards further nitropyrrolidines, nitropyrrolizines, and nitropyroles [10].

A sophisticated continuous flow instrumental setup was assembled and used for the uninterrupted synthesis and purification of several 1,2-pyrazoles [11]. Combining a Vapourtec flow system with a Uniqsis FlowSyn system and a MultiplexIR flow cell, a fully automated and software controllable flow system was established. As a second chemistry example, a Roush crotylation sequence was also demonstrated. Besides 1,2-pyrazoles, pyrimidines and flavones were synthesized in a similar fashion by the same group [12].

Two bicyclic-Δ^2-isoxazolines were synthesized as constrained analogues of procaine/procainamide and tested as inhibitors of DNA methyltransferase 1 [13]. The key step in the synthesis was the 1,3-dipolar cycloaddition of nitrile oxides to *N*-Boc-protected pyrroline. The nitrile oxides were formed in situ by base-induced dehydrochlorination of the intermediate chlorooximes. The low reactivity of the *N*-Boc-protected pyrroline and the low yields of the cycloaddition products obtained under batch conditions were positively affected when running the reaction under continuous flow conditions. The *N*-Boc-protected bicyclic-Δ^2-isoxazolines were generated in higher yields at 80 °C reaction temperature and only 10 min reaction time.

A series of 4-(pyrazol-1-yl)carboxanilides with inhibitory effect on canonical TRP-channels were synthesized using microwave and continuous flow technology [14]. An efficient three-step protocol comprising cyclocondensation of 4-nitrophenylhydrazine with appropriate 1,3-dicarbonyl building blocks, sequential nitro-reduction, and subsequent batch amidation with carboxylic acids provided the targeted heterocyclic carboxanilides. Alternatively, cyclocondensation of

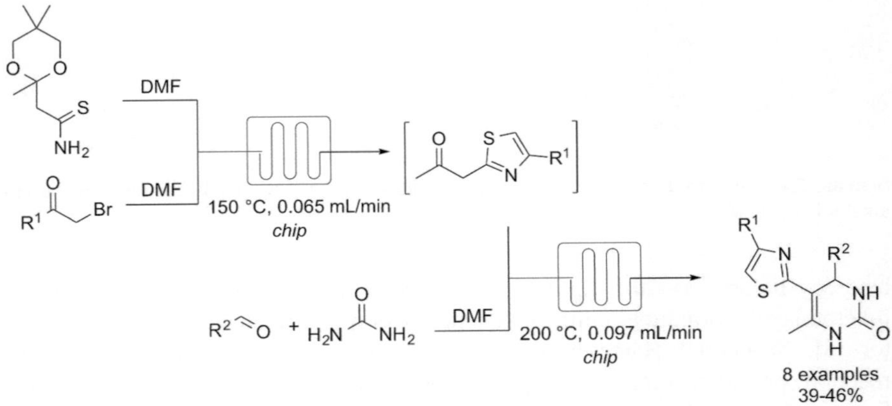

Scheme 7.5 Sequential Hantzsch dihydropyrimidine flow synthesis

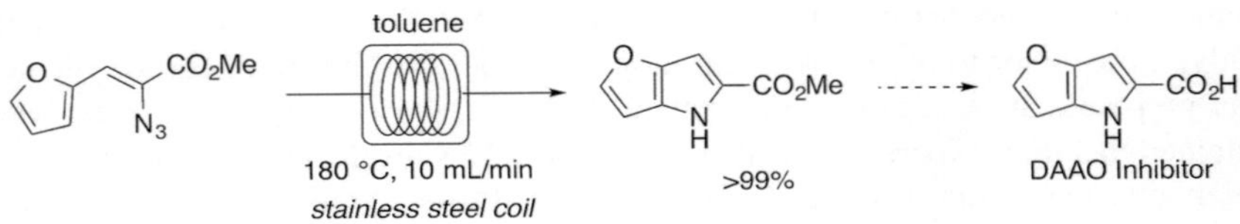

Scheme 7.6 Continuous flow thermolysis of azidoacrylates towards a DAAO inhibitor

4-bromophenylhydrazine with appropriate 1,3-dicarbonyl building blocks followed by a Buchwald-Hartwig amidation also resulted in 4-(pyrazol-1-yl)carboxanilides (Scheme 7.7). In a similar process, a small molecular library of thiazoles and pyrazoles was reported by Thompson et al. [15].

A microwave-to-flow method translation was applied to prepare 2-heptyl-3-hydroxyl-4(1*H*)-quinolone as a *Pseudomonas* quinolone signal (PQS). In a reaction of an α-chloro ketone with commercial anthranilic acid, the PQS compound was prepared at 220 °C reaction temperature in only 5 min reaction time in 57 % yield [16].

An efficient and safe flow process for the synthesis of aromatic and aliphatic diazoketones and their subsequent transformation into quinoxalines was recently reported [17]. Reacting various acid chlorides and $TMSCHN_2$ for 25 min reaction time at room temperature, the corresponding diazoketones were isolated in 46–98 % yields. The optimized procedure was implemented in the telescoped synthesis of quinoxalines (Scheme 7.8). The flow stream containing the prepared diazoketones was combined with a stream of 1,2-diaminobenzenes and the resulting reaction mixture passed over a supported copper catalyst followed by series of scavenger resigns at a temperature of 110 °C to deliver a selection of quinoxalines in 21–73 % yield.

A research group at GlaxoSmithKline (UK) exploited the benefits of continuous flow technology to prepare small amounts of pharmaceutically interesting

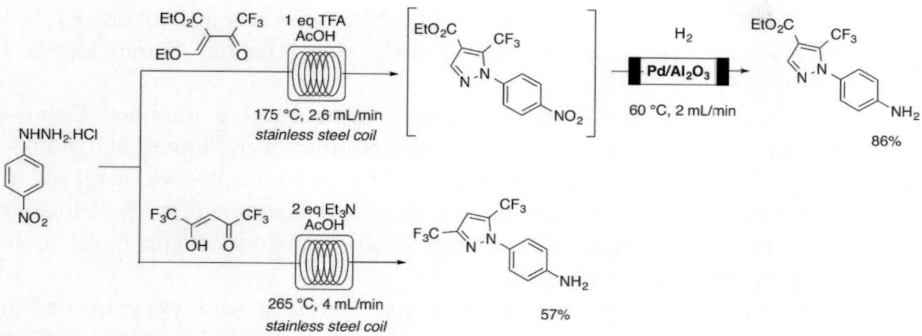

Scheme 7.7 Continuous flow steps in the synthesis of 4-(pyrazol-1-yl)carboxanilides

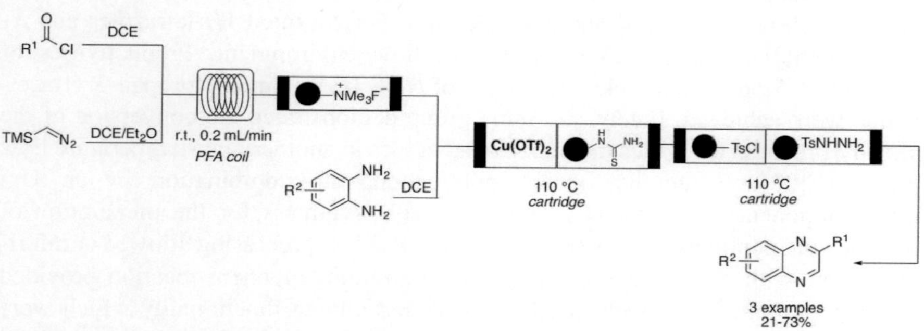

Scheme 7.8 Generation and utilization of diazoketones in the flow synthesis of quinoxalines

substituted indazoles [18]. By reacting several hydrazines with substituted 2-fluorobenzaldehydes, 2-fluorobenzonitriles, or 2-fluoromethylbenzoates at elevated temperature (>150 °C), small libraries of substituted indazoles were easily synthesized.

The Paal–Knorr synthesis of pyrroles was also examined under flow conditions [19]. Using a microstructured glass reactor with the optimized reaction temperature, stoichiometry, and reaction time, a synthetic throughput of 55.8 g/h was achieved.

Immobilized reagents and scavengers were used in the synthesis of the natural product *O*-methyl siphonazole [20]. Sequential Claisen condensation and Krapcho decarboxylation were the key steps for the successful synthesis. Using a combination of flow and batch methods, the desired product was obtained in a nine-step synthesis. The research group applied a very similar approach in the synthesis of (−)-hennoxazole in a follow-up study [21].

A library of 3-aminoindolizines and some aza-indolizines were prepared in an orthogonal modification of a heterocyclic core prepared in situ [22, 23]. All three steps were realized with the aid of an automated flow instrument. Reacting 5-substituted 2-bromopyridine with different alkynes provided the corresponding 3-aminoindolizines in up to 78 % yields. The optimized Sonogashira/cycloisomerization process was carried out at 180 °C within 7 min reaction time.

Similar heterocyclic compounds—imidazo[1,2-b]-pyridazines and imidazo[1,2-a] pyridines—were prepared by Ley and coworkers as potential Casein kinase I inhibitors [24, 25].

Tri- and tetrasubstituted imidazoles were obtained via a modified Debus–Radziszewski reaction under continuous flow conditions, using imines and ammonium acetate [26]. The envisaged synthesis of *N*-hexyl-imidazoles was achieved in *n*-pentanol at 120 °C and 48 min residence time with a maximum of 36 % yield after extraction. The obtained heterocycles were finally transformed into novel ionic liquids with superior chemical stability [27].

In another study, the very toxic and explosive hydrazoic acid was generated in situ under flow conditions either by mixing of acetic acid with aqueous sodium azide or by mixing neat trimethylsilyl azide upon mixing with methanol [28]. The stream containing the hydrazoic acid was subsequently reacted with organic nitriles or 2-oxazolines to generate the corresponding 5-substituted-1*H*-tetrazoles and *N*-(2-azidoethyl)acylamides in a continuous flow environment. Productivities of 18.9 g/h of 5-phenyltetrazole and 23 g/h of *N*-(1-azido-2-methylpropan-2-yl)acetamide were achieved. Lately, the same group demonstrated the conversion of the already-prepared tetrazoles into 1,3,4-oxadiazoles in another high-temperature flow protocol [29]. In another recent publication, the combination of an Ugi multicomponent reaction and a "click"-triazole synthesis for the preparation of macrocyclic peptidomimetics was reported [30]. Flow processing allowed comfortable work with malodorous isocyanides. The multicomponent reaction provided linear peptidomimetics with installed azide and alkyne functionality which were then quantitatively "clicked" to a macrocyclic triazole inside a copper coil in 25 min reaction time and 140 °C reaction temperature. A different approach was developed for the synthesis of 5-amino-2-aryl-2*H*-[1,2,3]-triazole-4-carbonitriles [31]. Reacting commercially available anilines and malononitrile in the presence of *tert*-butyl nitrite as the oxidant, 2-arylhydrazonomalononitrile were obtained at room temperature. In the next step, the 2-arylhydrazonomalononitriles were converted into 2-arylhydrazono-2-cyanoacetamidines by reaction with methanolic ammonia or secondary amines at 110 °C. The final oxidative cyclization was catalyzed by copper acetate at 140 °C reaction temperature in the presence of sodium periodate as the most effective oxidant (Scheme 7.9). A small library of 2-substituted-[1,2,3]-triazoles was finally prepared.

Employing ethyl isocyanoacetate produced in a flow process, the telescoped synthesis of 1,2,4-triazoles and pyrrolo-[1,2-c]-pyrimidine was accomplished [32]. Using *N*-formylglycine and triphosgene, nearly quantitative amounts of ethyl isocyanoacetate were generated at room temperature. Straightforward reaction with either in situ generated diazonium species or with pyrrole-2-carboxaldehyde delivered the final products (Scheme 7.10).

Small, drug-like 2-(1*H*-indol-3-yl)thiazoles were generated in an automated flow system [33]. The flow assembly of the molecules was based on a consecutive Hantzsch thiazole synthesis, deketalization, and Fischer indole synthesis. The reactions were performed using a chip microreactor setup at temperatures of

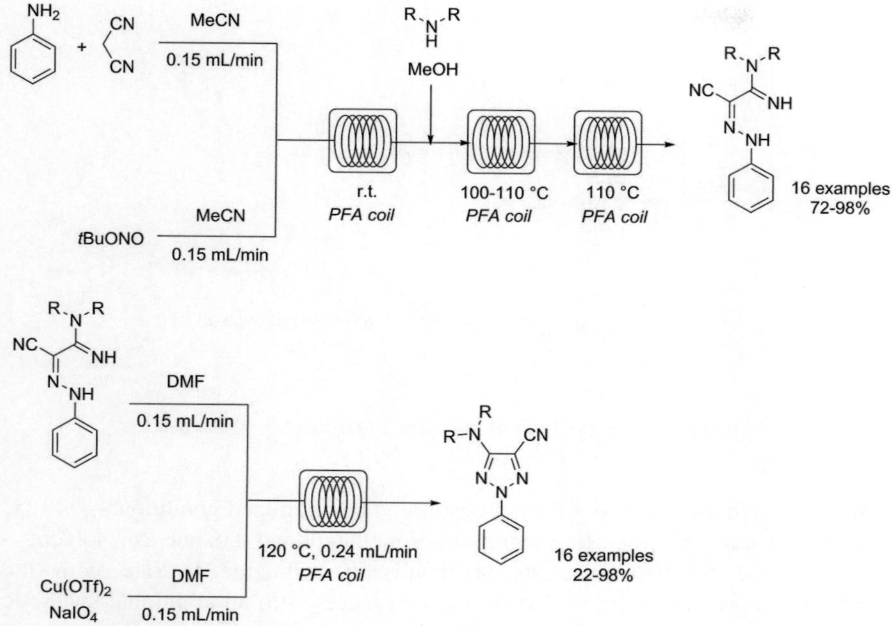

Scheme 7.9 Flow synthesis of 5-amino-2-aryl-2*H*-[1,2,3]-triazole-4-carbonitriles

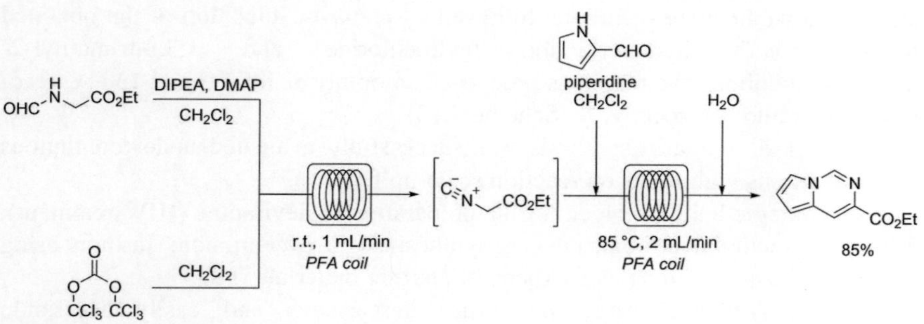

Scheme 7.10 Telescoped synthesis of pyrrolo-[1,2-c]-pyrimidine from N-formylglycine

150–200 °C, and reaction times of 3–10 min per single step were sufficient to provide the final 2-(1*H*-indol-3-yl)thiazoles (Scheme 7.11).

Substituted heteroaromatic compounds including pyridopyrimidinones and hydroxyquinolines were synthesized employing a high-temperature (>300 °C) and high-pressure (100–160 bar) continuous flow protocol via intramolecular thermal cyclization and benzannulation reactions of the Gould–Jacobs and Conrad–Limpach types [34].

A research group at Abbott (USA) developed a continuous flow method for the synthesis of diaminopyrazoles [35]. After an initial microwave optimization, the

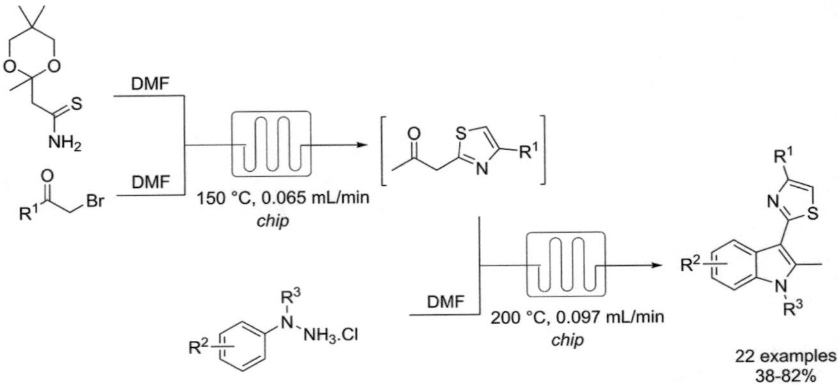

Scheme 7.11 Continuous flow synthesis of drug-like 2-(1*H*-indol-3-yl)-thiazoles

process was translated into a flow procedure. The optimized conditions—140 °C and 10 min reaction time using a mixture of methanol and dioxane as a solvent—delivered the target compounds upon mixing hydrazine hydrate with the corresponding malononitriles. Extending the process with an additional synthetic step allowed the synthesis of a pyrazolo-[1,5-α]-pyrimidine (Scheme 7.12).

Commercially relevant photochromic spirooxazines were prepared recently [36]. The two-step synthesis involved a copper-catalyzed addition of substituted anilines to naphthalene-1,2-dione followed by a *spiro*-cyclization of the obtained products with hydroxylamine hydrochloride and 1,3,3-trimethyl-2-methyleneindoline. The reactions proceeded smoothly at 140 °C and 120 °C reaction temperatures, respectively (Scheme 7.13).

The Perkin coumarin synthesis was successfully evaluated under continuous flow conditions with focus on reaction scale-up [37].

As a strategic building block for the preparation of nevirapine (HIV treatment), 2-bromo-4-methylnicotinonitrile was synthesized in a continuous fashion using malononitrile and acetone as inexpensive starting materials [38].

Another synthesis, based on rather inexpensive and easily accessible 6-aminouracil, formaldehyde, and amines, was disclosed by Stevens et al. recently [39]. The batch microwave synthesis of tricyclic 5,6,8,9-tetrahydro-4*H*,7*H*-2,5,6a,8,9a-pentaazaphenalene-1,3-diones was evaluated under flow conditions with focus on reaction scale-up. The automated flow device allowed productivity of 602 mg/h.

A heterogeneously catalyzed flow synthesis of γ-valerolactone from levulinic acid was achieved under hydrogen atmosphere and using various catalysts [40]. By reacting 2-aminopyrimidine with different aldehydes and isocyanides catalyzed by Zr^{4+} or Sc^{3+} Lewis acids, 3-aminoimidazo[1,2-a]-pyrimidines were synthesized via the Groebke–Blackburn tricomponent reaction. The rather mild reaction conditions (80 °C temperature, 50 min reaction time) provided almost exclusively the 3-amino isomers.

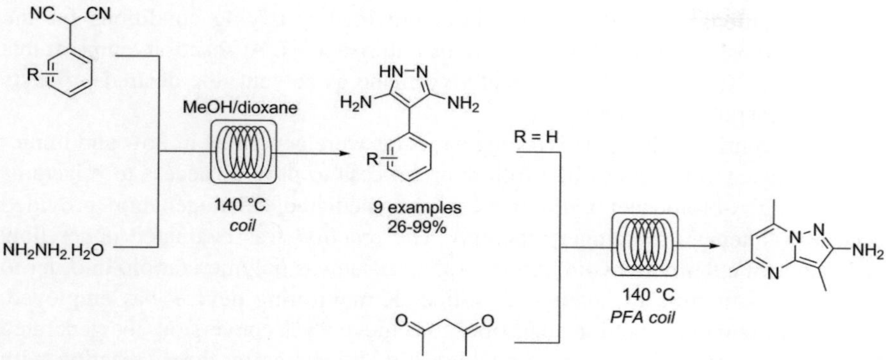

Scheme 7.12 Continuous flow method for the synthesis of diaminopyrazoles and a telescoped synthesis of a pyrazolo-[1,5-α]-pyrimidine

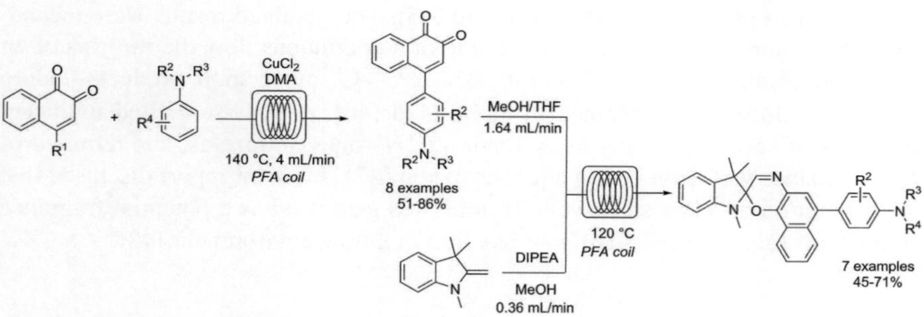

Scheme 7.13 Continuous flow synthesis of spirooxazines as photochromic dyes

A recent example of a continuous dibenzodiazepine synthesis was reported by Baxendale et al. [41]. A sequential S_NAr-reaction of 2-fluoronitroarenes with 2-aminobenzophenones and an intramolecular reductive amination provided two dibenzodiazepines in 51 % (120 mmol scale) and 68 % yield (50 mmol scale). In-line extraction was introduced in between the two reaction steps as a continuous purification procedure.

A number of tricyclic pyrrolo[1,2-a]-quinolines bearing phosphonate or phosphine oxide moieties were synthesized via an allene-based cascade in flow [42]. Transformation of propargylic alcohols into allenes in presence of chlorophosphine/phosphite reagents by a [2,3]-sigmatotropic rearrangement is the first step of the process. In situ trapping of the formed allene by a pyrrole ring terminates the process to deliver pyrrolo[1,2-a]-quinolines on a multigram scale.

Ley and coworkers developed a simple and rapid synthetic protocol to obtain valuable oxazolines and related oxazoles [43]. By mixing streams containing β-hydroxy amide and Deoxo-Fluor® at room temperature, followed by a basic quench of the excess HF and an in-line extraction, the expected oxazolines were obtained in >60 % yields. Benefiting from earlier experimental experience with

MnO_2-oxidations, a rapid reoptimization provided satisfying conditions for the oxidative transformation of the oxazolines into oxazoles. At reaction temperatures of either 40 °C or 100 °C and dimethoxyethane as solvent, the desired products were successfully generated.

Most recently, highly reactive ketene species were generated in flow and immediately reacted with imines in a follow-up process to provide access to β-lactams [44]. Using α-bromoacetyl bromides, a zinc-mediated dehalogenation provided access to ketenes (Staudinger reaction). The reaction was evaluated under flow conditions in attempt to avoid known problems such as polymerization. In order to follow the formation of ketene, an in-line IR monitoring device was employed. After optimizing the reaction conditions to achieve 98 % conversion, the generated ketene was used to prepare a small library of β-lactams by direct reaction with various imines. Good yields were obtained although with only low *cis/trans* selectivity.

Using a high-temperature and pressure conditions, few selected examples of a flow pyrolysis process were demonstrated [45]. The obtained results were mechanistically rationalized. In a single example of a continuous flow thermolysis of an enyne-carodiimide at 150 °C within 4 h, a C^2–C^6 cyclization product—indolo [2,3-b]quinolone—was obtained in 83 % yield [46]. A PEG-supported multistep synthesis of several 3,4-dihydropyrimidin-2(1*H*)-ones, tetrazoles, and tetrahydro-1,3-isoxazines was achieved in aqueous media [47]. A recent report disclosed that the generation of a key spirocyclic lactone was generated as a potential fragrance component using a Baylis–Hillman reaction in a flow environment [48].

7.2 Multistep Syntheses

A flow process for the multistep synthesis of the alkaloid (±)-oxomaritidine (Scheme 7.14) was established by the combination of multiple flow reactions using supported reagents, catalysts, and scavengers in tubular and chip reactors [49]. A combination of flow instruments from different vendors was needed to achieve the set goal in only 6 h lasting continuous process. The flow synthesis of *N,N*-diethyl-4-(3-fluorophenylpiperidin-4-ylidenemethyl)-benzamide, a potent δ-opioid receptor agonist developed by AstraZeneca, was optimized using a telescoped four-step concept (Scheme 7.15) [50]. Multiple cartridge-packed reagents and scavengers were used.

The developed platform was used by the research group for the synthesis of various other compounds with pharmacological significance or with natural origin [51–57].

A multistep approach towards the synthesis of an aminonaphthalene derivative as a key intermediate in the synthesis of duocarmycin prodrug for cancer treatment was described utilizing a microreactor. In most of the steps, similar or slightly better results as compared to corresponding batch experiments were obtained [58].

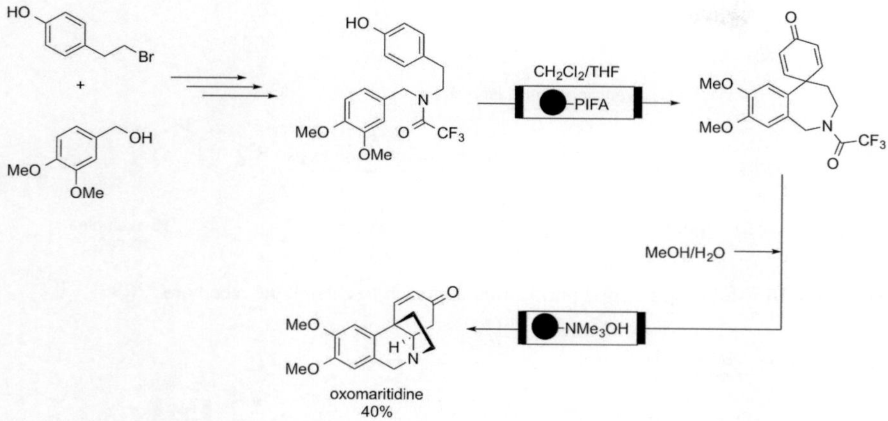

Scheme 7.14 Multistep all-flow synthesis of oxomaritidine

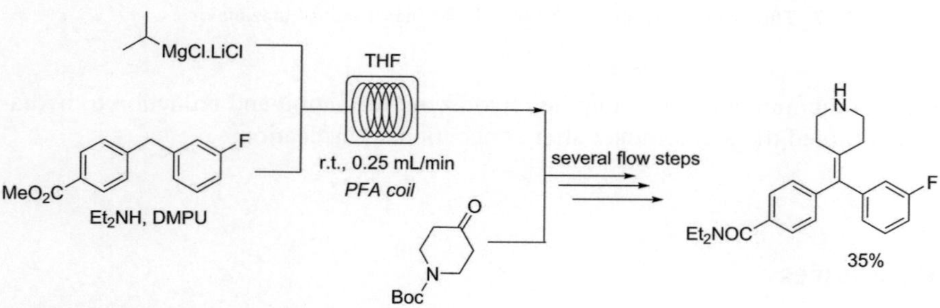

Scheme 7.15 Flow approach towards a potent δ-opioid receptor agonist

Following recent synthetic trends, a research team at Eli Lilly (UK) evaluated the performance of an automated flow system for the multiple step synthesis of (±)-fluoxetine [59]. The optimized protocol delivered the product in 86 % yield after purification by chromatography (4.8 mmol/h productivity).

Ligands for the chemokine receptor CCR8—a common target in various inflammatory and allergic conditions—were prepared using the concept of supported scavengers and reagents in a flow procedure [60]. Reaction of an amine with isocyanate, a reductive Cbz-deprotection, and an alkylation of a secondary amine were performed in a flow sequence to generate a 15-membered library of compounds (Scheme 7.16).

The total synthesis of (+)-dumetorine, (−)-seadmine and (+)-sedridine was accomplished using flow instrumentation. Five synthetic steps led to (+)-dumetorine with good yields and minimized the required efforts as compared to a batch synthesis [61].

In a three-step sequential flow synthesis, the biaryl unit of atazanavir (an HIV protease inhibitor) was prepared in 74 % overall yield (Scheme 7.17)

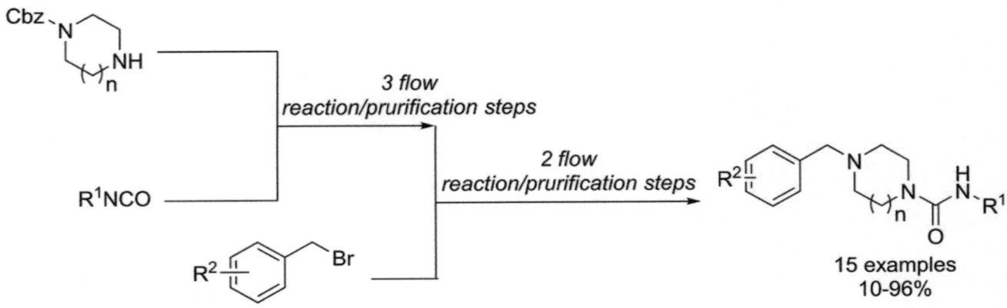

Scheme 7.16 Multistep reaction/purification approach to chemokine receptor CCR8

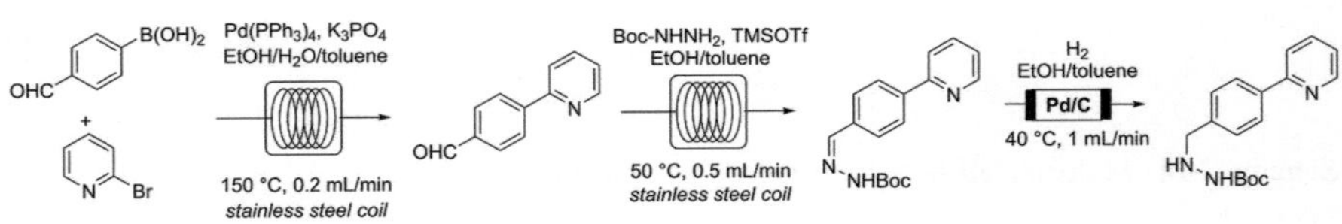

Scheme 7.17 Three-step flow approach towards the biaryl unit of atazanavir

[62]. Combining a Suzuki coupling, hydrazone formation and reduction to hydrazine afforded the pure product after conventional purification.

References

1. T. Schwalbe, D. Kadzimirsz, G. Jas, QSAR Comb. Sci. **24**, 758 (2005)
2. D.R.J. Acke, C.V. Stevens, Green Chem. **9**, 386 (2007)
3. I.R. Baxendale, S.V. Ley, C.D. Smith, L. Tamborini, A.-F. Voica, J. Comb. Chem. **10**, 851 (2008)
4. D.R.J. Acke, C.V. Stevens, B.I. Roman, Org. Process Res. Dev. **12**, 921 (2008)
5. M. Brasholz, B.A. Johnson, J.M. Macdonald, A. Polyzos, J. Tsanaktsidis, S. Saubern, A.B. Holmes, J.H. Ryan, Tetrahedron **66**, 6445 (2010)
6. M. Baumann, I.R. Baxendale, C. Kuratli, S.V. Ley, R.E. Martin, J. Schneider, ACS Comb. Chem. **13**, 405 (2011)
7. W.G. Devine, N.E. Leadbeater, ARKIVOC **v**, 127 (2011)
8. N. Pagano, A. Herath, N.D.P. Cosford, J. Flow Chem. **1**, 28 (2011)
9. A.G. O'Brien, F. Levesquea, P.H. Seeberger, Chem. Commun. **47**, 2688 (2011)
10. M. Baumann, I.R. Baxendale, A. Kirschning, S.V. Ley, J. Wegner, Heterocycles **82**, 1297 (2011)
11. H. Lange, C.F. Carter, M.D. Hopkin, A. Burke, J.G. Goode, I.R. Baxendale, S.V. Ley, Chem. Sci. **2**, 765 (2011)
12. I.R. Baxendale, S.C. Schou, J. Sedelmeier, S.V. Ley, Chem. Eur. J. **16**, 89 (2010)
13. S. Castellano, D. Kuck, M. Viviano, J. Yoo, F. Lopez-Vallejo, P. Conti, L. Tamborini, A. Pinto, J.L. Medina-Franco, G. Sbardella, J. Med. Chem. **54**, 7663 (2011)
14. D. Obermayer, T.N. Glasnov, C.O. Kappe, J. Org. Chem. **76**, 6657 (2011)
15. C.M. Thompson, J.L. Poole, J.L. Cross, I. Akritopoulou-Zanze, S.W. Djuric, Molecules **16**, 9161 (2011)

16. J.T. Hodgkinson, W.R.J.D. Galloway, S. Saraf, I.R. Baxendale, S.V. Ley, M. Ladlow, M. Welch, D.R. Spring, Org. Biomol. Chem. **9**, 57 (2011)
17. L.J. Martin, A.L. Marzinzik, S.V. Ley, I.R. Baxendale, Org. Lett. **13**, 320 (2011)
18. R.C. Wheeler, E. Baxter, I.B. Campbell, S.J.F. Macdonald, Org. Process Res. Dev. **15**, 565 (2011)
19. P.J. Nieuwland, R. Segers, K. Koch, J.C.M. van Hest, F.P.J.T. Rutjes, Org. Process Res. Dev. **15**, 783 (2011)
20. M. Baumann, I.R. Baxendale, M. Brasholz, J.J. Hayward, S.V. Ley, N. Nikbin, Synlett **2011**, 1375 (2011)
21. A. Fernandez, Z.G. Levine, M. Baumann, S. Sulzer-Mosse, C. Sparr, S. Schläger, A. Metzger, I.R. Baxendale, S.V. Ley, Synlett **24**, 514 (2013)
22. P.P. Lange, K. James, ACS Comb. Sci. **14**, 570 (2012)
23. P.P. Lange, A.R. Bogdan, K. James, Adv. Synth. Catal. **354**, 2373 (2012)
24. F. Venturoni, N. Nikbin, S.V. Ley, I.R. Baxendale, Org. Biomol. Chem. **8**, 1798 (2010)
25. L. Guetzoyan, N. Nikbin, I.R. Baxendale, S.V. Ley, Chem. Sci. **4**, 764 (2013)
26. D.R.J. Acke, R.V.A. Orru, C.V. Stevens, QSAR & Comb. Chem. **25**, 474 (2006)
27. C. Maton, N. De Vos, B.I. Roman, E. Vanecht, N.R. Brooks, K. Binnemans, S. Schaltin, J. Fransaer, C.V. Stevens, ChemPhysChem **13**, 3146 (2012)
28. B. Gutmann, D. Obermayer, J.-P. Roduit, D.M. Roberge, C.O. Kappe, J. Flow Chem. **1**, 8 (2012)
29. B. Reichart, C.O. Kappe, Tetrahedron Lett. **53**, 952 (2012)
30. C.E.M. Salvador, B. Pieber, P.M. Neu, A. Torvisco, C.K.Z. Andrade, C.O. Kappe, J. Org. Chem. **80**, 4590 (2015)
31. J. Jacq, P. Pasau, Chem. Eur. J. **20**, 12223 (2014)
32. M. Baumann, A.M.R. Garciaa, I.R. Baxendale, Org. Biomol. Chem. **13**, 4231 (2015)
33. N. Pagano, M.L. Heil, N.D.P. Cosford, Synthesis **44**, 2537 (2012)
34. L. Lengyel, T.Z. Nagy, G. Sipos, R. Jones, G. Dorman, L. Ürge, F. Darvas, Tetrahedron Lett. **53**, 738 (2012)
35. N.S. Wilson, A.T. Osuma, J.A. Van Camp, X. Xu, Tetrahedron Lett. **53**, 4498 (2012)
36. M. York, A. Edenharter, Aust. J. Chem. **66**, 172 (2013)
37. A. Chen, X. li, Y. Zhou, L. Huang, Z. Fang, H. Gan, K. Guo, Adv. Mat. Res. **781–784**, 936 (2013)
38. A.R. Longstreet, S.M. Opalka, B.S. Campbell, B.F. Gupton, D.T. McQuade, Beilstein J. Org. Chem. **9**, 2570 (2013)
39. D. Garcia, M. Jatczak, K. Muylaert, L.M. De Coen, C.V. Stevens, Eur. J. Org. Chem. **1732** (2013)
40. A.J.E. Butler, M.J. Thompson, P.J. Maydom, J.A. Newby, K. Guo, H. Adams, B. Chen, J. Org. Chem. **79**, 10196 (2014)
41. M. Baumann, I.R. Baxendale, C.H. Hornung, S.V. Ley, M.V. Rojo, K.A. Roper, Molecules **19**, 9736 (2014)
42. M. Baumann, I.R. Baxendale, J. Org. Chem. **80**, 10806 (2015)
43. S. Glöckner, D.N. Tran, R.J. Ingham, S. Fenner, Z.E. Wilson, C. Battilocchio, S.V. Ley, Org. Biomol. Chem. **13**, 207 (2015)
44. A. Hafner, S. Ley, Synlett **26**, 1470 (2015)
45. M. Schmittel, J.-P. Steffen, D. Rodriguez, B. Engelen, E. Neumann, M.E. Cinar, J. Org. Chem. **73**, 3005 (2008)
46. D. Cantillo, H. Sheibani, C.O. Kappe, J. Org. Chem. **77**, 2463 (2012)
47. N. Prosa, R. Turgis, R. Piccardi, M.-C. Scherrmann, Eur. J. Org. Chem. **2188** (2012)
48. I.R. Baxendale, Chem. Eng. Technol. **38**, 1713 (2015)
49. I.R. Baxendale, J. Deeley, C.M. Griffiths-Jones, S.V. Ley, S. Saaby, G.K. Tranmer, Chemm. Commun. **2006**, 2566 (2006)
50. Z. Qian, I.R. Baxendale, S.V. Ley, Chem. Eur. J. **16**, 12342 (2010)
51. Z. Qian, I.R. Baxendale, S.V. Ley, Synlett **21**, 505 (2010)

52. R.J. Ingham, E. Riva, N. Nikbin, I.R. Baxendale, S.V. Ley, Org. Lett. **14**, 3920 (2012)
53. P.R.D. Murray, D.L. Browne, J.C. Pastre, C. Butters, D. Guthrie, S.V. Ley, Org. Process Res. Dev. **17**, 1192 (2013)
54. S. Newton, C.F. Carter, C. M. Pearson, L. de C. Alves, H. Lange, P. Thansandote, S.V. Ley, Angew. Chem. Int. Ed. **53**, 4915 (2014)
55. R.J. Ingham, C. Battilocchio, J.M. Hawkins, S.V. Ley, Beilstein J. Org. Chem. **10**, 641 (2014)
56. R.J. Ingham, C. Battilocchio, D.E. Fitzpatrick, E. Sliwinski, J.M. Hawkins, S.V. Ley, Angew. Chem. Int. Ed. **54**, 144 (2015)
57. S.-H. Lau, A. Galvan, R.R. Merchant, C. Battilocchio, J.A. Souto, M.B. Berry, S.V. Ley, Org. Lett. **17**, 3218 (2015)
58. L. F. Tietze, D. Liu, ARKIVOC (**viii**), 193 (2008)
59. B. Ahmed-Omer, A.J. Sanderson, Org. Biomol. Chem. **9**, 3854 (2011)
60. T.P. Petersen, A. Ritzen, T. Ulven, Org. Lett. **11**, 5134 (2009)
61. E. Riva, A. Rencurosi, S. Gagliardi, D. Passarella, M. Martinelli, Chem. Eur. J. **17**, 6221 (2011)
62. L. Dalla-Vechia, B. Reichart, T. Glasnov, L.S.M. Miranda, C.O. Kappe, R.O.M.A. de Souza, Org. Bimol. Chem. **11**, 6806 (2013)

Chapter 8
Organic Synthesis in Dedicated Continuous Flow Systems

Synthesis of Radiolabeled Compounds

8.1 [^{18}F]-Labeled PET Radiotracers

The synthesis of [^{18}F]-fallypride in 0.5–1.5 mCi doses for micro-PET studies of brain dopamine subtype-2-receptors was produced in a microflow device from Advion [1]. Optimal conditions were rapidly scouted (reaction temperature, stoichiometry, flow rate, etc.) to deliver a reproducible decay-corrected radiochemical yield (RCY) of up to 88 % via a tosyl/[^{18}F]-fluoride ion exchange. The group investigated also the flow synthesis of sensitive [^{18}F]-labeled ligands for PET imaging of brain peripheral benzodiazepine receptors (Scheme 8.1) [2]. The final step in the synthesis of the 2-phenoxyanilides was the bromo/[^{18}F]-fluoride ion exchange under flow conditions at 100 °C for 10 min to obtain a 97 % RCY.

Generating another derivative required harsher conditions at 160 °C reaction temperature. Treating xenon difluoride with cyclotron-produced [^{18}F]-ions provided ^{18}F-labeled xenon difluoride. At elevated temperatures (>85 °C), the compound was produced in only 95 s on a microfluidic platform [3]. A 10 min run furnished sufficient amounts of the labeled fluoride to be subsequently used for the ^{18}F-fluorination of 1-((trimethylsilyl)oxy)cyclohexene (Scheme 8.2) and fluorene. The chloro/[^{18}F]-fluoride ion exchange under flow conditions was used for the generation of the cannabinoid receptor ligand CB41 as well as some precursors—ethylditosylate (EtDT) and propylditosylate (PrDT) [4, 5].

The group also presented the fully automated synthesis of [^{18}F]-MEL050 (melatonin) using the same chemical manipulation [6]. Lately, the synthesis of [^{18}F]-fluoromisonidazole was achieved by applying the tosyl//[^{18}F]-fluoride ion exchange method under flow conditions [7]. Labeling of the starting 1-(2′-nitro-1′-imidazolyl)-2-*O*-tetrahydropyranyl-3-*O*-toluenesulfonylpropanediol was followed by acidic hydrolysis with 2 N HCl to provide the final labeled product after HPLC-purification. The radiosynthesis of 11-[^{18}F]-fluoroundecyltriphenylphosphonium (MitoF) as a potential mitochondria-specific PET radiotracer was reported by applying the tosyl//[^{18}F]-fluoride ion exchange [8].

T. Glasnov, *Continuous-Flow Chemistry in the Research Laboratory*,
DOI 10.1007/978-3-319-32196-7_8

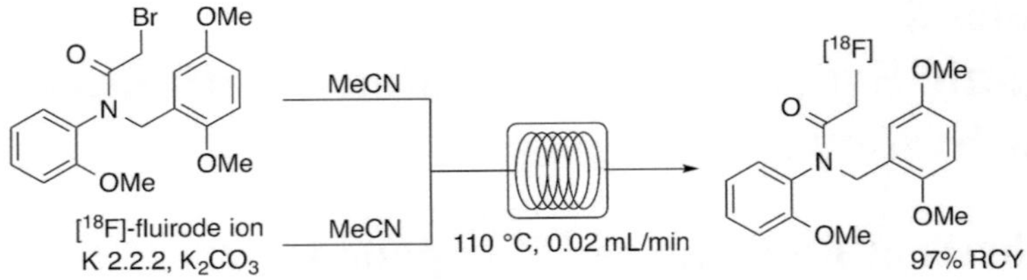

Scheme 8.1 Bromo/[^{18}F]-fluoride ion exchange under flow conditions

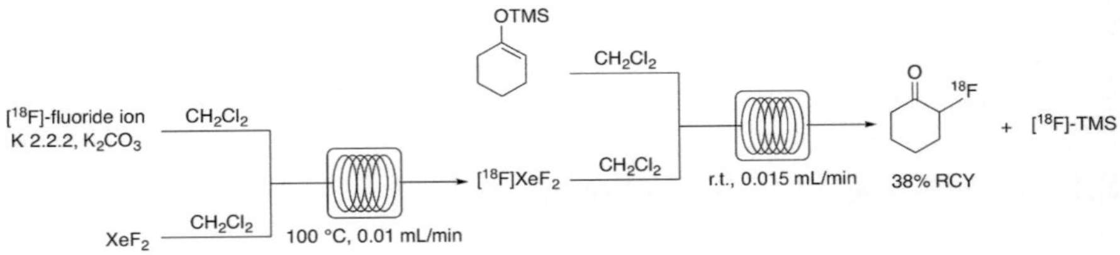

Scheme 8.2 Generation and usage of [^{18}F]XeF_2 for the fluorination of 1-((trimethylsilyl)oxy) cyclohexene in flow

Pike et al. developed another microfluidic protocol for the preparation of *ortho*-substituted [^{18}F]-fluoroarenes. Reacting cyclotron-generated [^{18}F]-ions (100–150 mCi) with *ortho*-substituted aryliodonium salts at 140–190 °C rapidly provided the desired radiotracers in 51–85 % RCY. The *ortho*-effect of various substituents on the product distribution was also evaluated [9]. In a later report, the same approach was conveniently adopted for the synthesis of 2- and 3-[^{18}F]-fluorohalopyridines as well as for some [^{18}F]-labeled aromatic "click" synthons [10, 11]. Instead of iodonium salts, a S_NAr substitution with a 4-trimethylammoniumbenzaldehyde triflate was adopted for the generation of 4-[^{18}F]-fluorobenzaldehyde [12]. Aryliodonium salt was also used by Caroll et al. in the phase-transfer microfluidic generation of [^{18}F]-fluoroarenes [13]. Additionally, nitroarene and [^{19}F]-fluoroarene were also examined as possible precursors. The radiofluorodenitration was also applied for the high-temperature synthesis of [^{18}F]-altanserin [14]. In the same way, various substituted [^{18}F]-fluoroarenes were synthesized by the group of Kabalka [15]. *Ortho*-, *meta*-, and *para*-substituted nitroarenes were tested as substrates and afforded varying yields in the range of 35–97 % and within reaction times of less than 5 min. Collier, Vasdev, and coworkers successfully used the nitro/[^{18}F]-exchange method to prepare 3-[^{18}F]-fluoro-5-[(pyridin-3-yl)ethynyl] benzonitrile ([^{18}F]-FPEB)—a PET-tracer for imaging the metabotropic glutamate receptor subtype type 5 (mGluR5) [16]. Utilizing a microfluidic system, the high-temperature process (215 °C) provided the product in >95 % radiochemical purity. The authors used the same high-temperature approach to obtain and test another PET-tracer—7-(6-[^{18}F]-fluoropyridin-3-yl)-5H-pyrido[4,3-b]indole ([^{18}F]-T807) [17]. In the latest report on the topic, the group developed a flow labeling protocol based on the use of spirocyclic hypervalent iodine(III) ylides instead of iodonium salts [18]. The new process was successfully applied for the synthesis of 3-[^{18}F]-fluoro-5-[(pyridin-3-yl)ethynyl]benzonitrile ([^{18}F]-FPEB) and for a routinely used building block for click-radiochemistry, 4-[^{18}F]-fluorobenzyl azide.

An improved synthesis of 2′-[^{18}F]-fluoro-2′-deoxy-1-β-D-arabinofuranosyl-5-iodouracil ([^{18}F]-FIAU) was disclosed by Lewis et al. in 2010 [19]. Nucleophilic fluorination of the sugar was successfully accomplished in flow and followed by a conventional trimethylsilyl trifluoromethanesulfonate (TMSOTf)-catalyzed coupling of 2-deoxy-2-[^{18}F]-fluoro-1,3,5-tri-*O*-benzoyl-D-arabinofuranose with 2,4-bis(trimethylsilyloxy)-5-iodouracil to yield the protected dibenzoyl-[^{18}F]-FIAU. After deprotection with sodium methoxide, the isomeric product mixture was purified by HPLC to provide the β-anomer of [^{18}F]-FIAU). Another sugar PET-reagent was synthesized by Wuest et al.—1-(5-deoxy-5-fluoro-α-D-arabinofuranosyl)-2-nitroimidazole ([^{18}F]-FAZA). A nucleophilic tosylate/[^{18}F]-fluoride exchange and a deprotection step were included in the two-step synthesis [20]. Different reaction parameters were studied before RCYs of 40–63 % were reached. The obtained [^{18}F]-FAZA was used in dynamic small animal PET studies on EMT-6 tumor-bearing BALB/c mice. The group reported also an economical and versatile synthesis of *O*-(2-[^{18}F]-fluoroethyl)-L-tyrosine ([^{18}F]FET) using the same microfluidic device [21].

The microfluidic preparation of [5-(2-[^{18}F]-fluoroethyl)2,4-diethyl-3-(ethylsulfanylcarbonyl)-6-phenylpyridine-5-carboxylate] ([^{18}F]-FE@SUPPY) and [5-ethyl-2,4-diethyl-3-((2-[^{18}F]-fluoroethyl)sulfanylcarbonyl)-6-phenylpyridine-5 carboxylate] ([^{18}F]-FE@SUPPY:2) was recently reported by Wadsak et al. (Scheme 8.3) [22]. A significant increase in RCY as compared to the batch method was observed. The same research laboratory reported the synthesis of [^{18}F]-FE@SNAP as a PET-tracer for the melanin concentrating hormone receptor 1 [23]. [^{18}F]-Fluoroethylation was attempted using various [^{18}F]-fluoroalkylated synthons. However, only the direct [^{18}F]-fluorination approach using a tosylated precursor inside a flow-through microreactor was successful, affording [^{18}F]-FE@SNAP in 44 % yield at 170 °C reaction temperature.

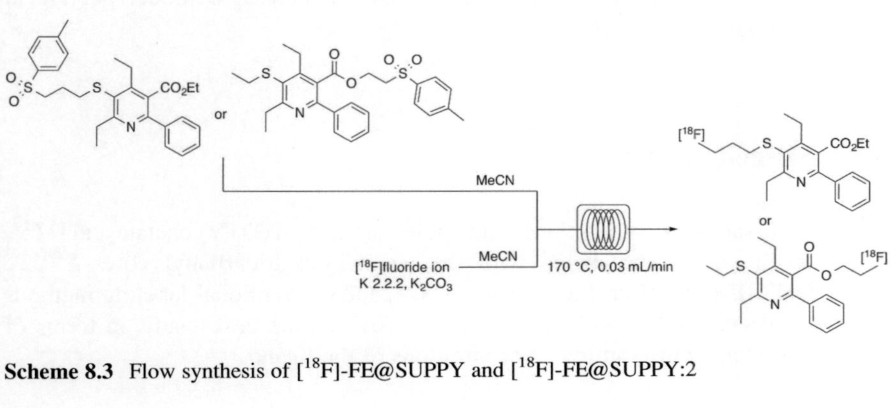

Scheme 8.3 Flow synthesis of [^{18}F]-FE@SUPPY and [^{18}F]-FE@SUPPY:2

The oxidative fluorination of 4-*tert*-butylphenols with no-carrier-added [^{18}F]-fluoride was optimized to deliver a library of 4-[^{18}F]-fluorophenols [24]. The reported one-pot protocol could be simply transferred to a flow device (room temperature, 15 μL/min flow rate), and 18 % RCY of 4-[^{18}F]-fluorophenol were obtained as a "proof-of-concept."

Using a tosylate precursor, [^{18}F]-FPMA (2-(5-[^{18}F]-fluoro-pentyl)-2-methyl malonic acid) was synthesized in a microfluidic device in about 50 min total synthesis time [25]. The obtained product was tested on a mice-model for biodistribution and blood clearance.

Employing the NanoTek microfluidic system, the synthesis of various aryl [^{18}F]-sulfonyl fluorides from the corresponding sulfonyl chlorides was achieved at a reaction temperature of 100 °C in only 2 min reaction time [26]. Subsequent stability evaluation was also performed, showing >97 % of intact products after 3 h at room temperature. Very recently the authors disclosed an extended and extremely detailed, step-by-step procedure for the microfluidic preparation of [^{18}F]-MEL050 (melatonin) as a sample PET-tracer compound [27].

[^{18}F]-Labeled phosphopeptide-cell-penetrating peptide dimers were synthesized for the study and elucidation of various signal transduction pathways [28]. For the radiolabeling process, *N*-succinimidyl 4-[^{18}F]-fluorobenzoate was attached to the N-terminal end of the phosphopeptide part of the dimer. The microfluidic bioconjugation provided better radiochemical yields and higher purity of the labeled product as compared to the conventional synthesis. In a follow-up report by the same group, a different approach to generate [^{18}F]-labeled peptides was applied [29]. Using the triarylsulfonium moiety as leaving group, a S_NAr-process was adopted. Working under flow conditions, the optimal reaction temperature of 70–80 °C assured a labeling efficiency of up to 90 % as compared to the maximum 20 % obtained in batch. Further efforts by the group resulted in optimized flow procedures for the preparation of further [^{18}F]-labeled tracers with peptide structure—[^{18}F]-FDG-TATE and [^{18}F]-FDG-MHO-GSH [30]. In a two-step uninterrupted flow process, Sutcliffe and coworkers managed to generate a model [^{18}F]-labeled peptide—[^{18}F]-F-Py-YGGFL [31]. As a prosthetic group, 6-[^{18}F]-fluoronicotinic acid 2,3,5,6-tetrafluorophenyl ester ([18 F]F-Py-TFP) was prepared and subsequently attached to the NH_2-Tyr-Gly-Gly-Phe-Leu model peptide at 50 °C with radiolabeling yields of 28 %.

8.2 Miscellaneous

In a comparison study, bifunctional dithiazole valeric acid (DTV) chelate and DTV-AHx-insulin were radiolabeled with the technetium tricarbonyl core—[^{99m}Tc $(CO)_3(OH_2)_3]^+$ [32]. Microfluidic, microwave, and conventional labeling methods were evaluated. The microfluidic approach provided the best results in terms of yields (40 %) and purity at low concentrations of the ligand.

$[^{11}C]$-Carbon monoxide is another important labeling reagent. However, the low solubility of the gaseous reagent as well as the short radioactive half-life of carbon-11 (20 min) has restricted the use of this radiolabeling reagent so far. Nevertheless, microfluidic technology was shown to be the appropriate processing technique for liquid-$[^{11}C]$-CO chemistry and enabled rapid reaction optimization with small quantities of reagents. The neuropeptide Y Y5 receptor antagonist $[^{11}C]$-*trans*-N-[5-(2-flurophenyl)-2-pyrimidinyl]-3-oxospiro[5-azaisobenzofurane-1(3H),1-′-cyclohexane]-4′-carboxamide ($[^{11}C]$-MK-0233) was successfully prepared via a palladium-catalyzed $[^{11}C]$-carbonylation at 160 °C reaction temperature and 31 min residence time in 81 % radiochemical purity [33]. In another study on a general set of $[^{11}C]$-methylation reactions, the generation and application of $[^{11}C]$-CH_3I and $[^{11}C]$-CH_3OTf as $[^{11}C]$-methylation reagent was explored [34].

A microfluidic procedure for the preparation of *S*-$[^{13}N]$-nitrosothiols and ^{13}N-labeled azo-compounds was also recently developed. The radiochemical conversions were in the range of 54–99 % [35].

References

1. S. Lu, A.M. Giamis, V.W. Pike, Curr. Radiopharm. **2**, 49 (2009)
2. E. Briard, S.S. Zoghbi, F.G. Simeon, M. Imaizumi, J.P. Gourley, H.U. Shetty, S. Lu, M. Fujita, R.B. Innis, V.W. Pike, J. Med. Chem. **52**, 688 (2009)
3. S. Lu, V.W. Pike, J. Fluor. Chem. **131**, 1032 (2010)
4. G. Pascali, G. Mazzone, G. Saccomanni, C. Manera, P.A. Salvadori, Nucl. Med. Biol. **37**, 547 (2010)
5. G. Pascali, G. Nannavecchia, S. Pitzianti, P.A. Salvadori, Nucl. Med. Biol. **38**, 637 (2011)
6. L. Matesic, A. Kallinen, N.A. Wyatt, T.Q. Pham, I. Greguric, G. Pascali, Aust. J. Chem. **68**, 69 (2015)
7. M.-Q. Zheng, L. Collier, F. Bois, O.J. Kelada, K. Hammond, J. Ropchan, M.R. Akula, D.J. Carlson, G.W. Kabalka, Y. Huang, Nucl. Med. Biol. **42**, 578 (2015)
8. L. Li, L. Brichard, L. Larsen, D.K. Menon, R.A.J. Smith, M.P. Murphy, F.I. Aigbirhio, J. Label. Comp. Radiopharm. **56**, 717 (2013)
9. J.-H. Chun, S. Lu, Y.-S. Lee, V.W. Pike, J. Org. Chem. **75**, 3333 (2010)
10. J.-H. Chun, V.W. Pike, Chem. Commun. **48**, 9921 (2012)
11. J.-H. Chun, V.W. Pike, Eur. J. Org. Chem. **2012**, 4541 (2012)
12. K. Dahl, M. Schou, C. Halldin, J. Label. Comp. Radiopharm. **55**, 455 (2012)
13. C.D. Reed, G.G. Launay, M.A. Carroll, J. Fluor. Chem. **143**, 231 (2012)
14. J. Ungersboeck, S. Richter, L. Collier, M. Mitterhauser, G. Karanikas, R. Lanzenberger, R. Dudczak, W. Wadsak, Nucl. Med. Biol. **39**, 1087 (2012)
15. T.M. Moore, M.R. Akula, L. Collier, G.W. Kabalka, Appl. Radiat. Isot. **71**, 47 (2013)
16. S.H. Liang, D.L. Yokell, R.N. Jackson, P.A. Rice, R. Callahan, K.A. Johnson, D. Alagille, G. Tamagnan, T.L. Collier, N. Vasdev, MedChemComm **5**, 432 (2014)
17. S.H. Liang, D.L. Yokell, M.D. Normandin, R.A. Rice, R.N. Jackson, T.M. Shoup, T.J. Brady, G. El Fakhri, T.L. Collier, N. Vasdev, Mol. Imag. **13**, 1 (2014)
18. S. Calderwood, T.L. Collier, V. Gouverneur, S.H. Liang, N. Vasdev, J. Fluor. Chem. **178**, 249 (2015)
19. H. Anderson, N. Pillarsetty, M. Cantorias, J.S. Lewis, Nucl. Med. Biol. **37**, 439 (2010)
20. V.R. Bouvet, M. Wuest, L.I. Wiebe, F. Wuest, Nucl. Med. Biol. **38**, 235 (2011)

21. V.R. Bouvet, M. Wuest, P.-H. Tam, M. Wang, F. Wuest, Bioorg. Med. Chem. Lett. **22**, 2291 (2012)
22. J. Ungersboeck, C. Philippe, L.-K. Mien, D. Haeusler, K. Shanab, R. Lanzenberger, H. Spreitzer, B.K. Keppler, R. Dudczak, K. Kletter, M. Mitterhauser, W. Wadsak, Nucl. Med. Biol. **38**, 427 (2011)
23. C. Philippe, J. Ungersboeck, E. Schirmer, M. Zdravkovic, L. Nics, M. Zeilinger, K. Shanab, R. Lanzenberger, G. Karanikas, H. Spreitzer, H. Viernstein, M. Mitterhauser, W. Wadsak, Bioorg. Med. Chem. **20**, 5936 (2012)
24. Z. Gao, Y.H. Lim, M. Tredwell, L. Li, S. Verhoog, M. Hopkinson, W. Kaluza, T.L. Collier, J. Passchier, M. Huiban, V. Gouverneur, Angew. Chem. Int. Ed. **51**, 6733 (2012)
25. G.K. Dewkar, G. Sundaresan, N. Lamichhane, J. Hirsch, C. Thadigiri, T. Collier, M.C.T. Hartman, G. Vaidyanthan, J. Zweit, J. Label. Comp. Radiopharm. **56**, 289 (2013)
26. L. Matesic, N.A. Wyatt, B.H. Fraser, M.P. Roberts, T.Q. Pham, I. Greguric, J. Org. Chem. **78**, 11262 (2013)
27. G. Pascali, L. Matesic, T.L. Collier, N. Wyatt, B.H. Fraser, T.Q. Pham, P.A. Salvadori, I. Greguric, Nat. Prot. **9**, 2017 (2014)
28. S. Richter, V. Bouvet, M. Wuest, R. Bergmann, J. Steinbach, J. Pietzsch, I. Neundorf, F. Wuest, Nucl. Med. Biol. **39**, 1202 (2012)
29. S.V. Selivanova, L. Mu, J. Ungersboeck, T. Stellfeld, S.M. Ametamey, R. Schiblia, W. Wadsak, Org. Biomol. Chem. **10**, 3871 (2012)
30. V.R. Bouvet, F. Wuest, Lab Chip **13**, 4290 (2013)
31. R.C. Cumming, D.E. Olbergb, J.L. Sutcliffe, RSC Adv. **4**, 49529 (2014)
32. R.W. Simms, P.W. Causey, D.M. Weaver, C. Sundararajan, K.A. Stephenson, J.F. Valliant, J. Label. Comp. Radiopharm. **55**, 18 (2012)
33. S. Kealey, C. Plisson, T.L. Collier, N.J. Long, S.M. Husbands, L. Martarello, A.D. Gee, Org. Biomol. Chem. **9**, 3313 (2011)
34. J. Ungersboeck, C. Philippe, D. Haeusler, M. Mitterhauser, R. Lanzenberger, R. Dudczak, W. Wadsak, Appl. Radiat. Isot. **70**, 2615 (2012)
35. V. Gaja, V. Gomez-Vallejo, M. Cuadrado-Tejedor, J.I. Borrell, J. Llop, J. Label. Comp. Radiopharm. **55**, 332 (2012)

Chapter 9
Organic Synthesis in Dedicated Continuous Flow Systems

Enzymatic Reactions

9.1 Enzymatic Esterification and Acetylation

The vast majority of enzymatic processes are currently realized in various batch reactors. Nevertheless, a number of reports already describe several continuous flow processes in the area of enzymatic transformations.

Poppe et al. used commercially available lipases (immobilized or lyophilized powder forms) to achieve an enantiomerically selective acetylation of racemic 1-phenylethanol, 1-cyclohexylethanol, and 1-phenylpropan-2-ol [1]. The lipase preparations were filled into stainless steel cartridges which were then mounted to an X-Cube flow system. High enantiomeric excess ($E > 100$) were achieved using the described setup, while at the same time virtually no difference between batch and continuous flow experiments could be observed. The effects of temperature (0–60 °C) and pressure (1–120 bar) on the kinetics of the reaction were also evaluated.

A similar concept was used to esterify stearic acid with (*R*,*S*)-1,2-isopropylidene glycerol [2]. Packed cartridges with immobilized lipase from *Rhizomucor miehei* were used at 40–60 °C reaction temperature, 10 bar pressure, and 0.1–3 ml/min flow rate (several seconds residence time) to achieve the desired esterification. A response surface analysis was used to optimize the reaction conditions that finally resulted in a 95 % conversion of the substrate in only 40 s of reaction time. The process was later on extended to a three-step chemo-enzymatic cascade to prepare 1-monoacylglycerols [3] and to evaluate the efficiency of microemulsion-based organogels as lipase-supports [4]. A virtually identical process was used to determine the performance of another enzyme—Lecitase Ultra [5]. The enzyme was immobilized in several different ways before being tested for activity. Importantly, the leaching of the protein from the packed-bed columns was also examined—an essential factor for the catalyst lifetime and its reusability. The group used the same lipase from *Rhizomucor miehei* to prepare secondary glucose esters with different fatty acids. Residence time of 24 min was required for the reaction to reach full

T. Glasnov, *Continuous-Flow Chemistry in the Research Laboratory*,
DOI 10.1007/978-3-319-32196-7_9

conversion [6]. In yet another study, the valorization of fatty acid waste from the food industry via esterification under flow and batch conditions was assessed [7]. Three different options were tested—free CAL-B lipase, silica-encapsulated CAL-B IMOBCB, and Novozyme 435—with CAL-B IMOBCB giving the best results. In another esterification process, isoamyl oleate was synthesized as a biolubricant [8]. The optimized enzymatic process based on Novozyme 435 could be run uninterrupted for 144 h.

With the idea of preparing a chiral ionic liquid, the kinetic resolution of 6,7-dihydro-5*H*-pyrrolo[1,2-a]imidazol-7-ol was performed under continuous flow conditions with the help of Novozyme 435 and isopropenyl acetate (Scheme 9.1) [9]. The operationally simple procedure was performed at 35 °C in acetonitrile as the solvent of choice. Immobilized Novozyme 435 was utilized also for the kinetic resolution of (±)-1,3,6-tri-*O*-benzyl-*myo*-inositol—a precursor for *myo*-inositol phosphates [10]. Straightforward alcoholysis with vinyl acetate in *tert*-butyl methyl ether (TBME) (1:10) at 50 °C afforded 50 % conversion and >99 % *ee* in only 3 min residence time. The packed enzyme was reused in nine consecutive runs before deterioration started.

Vinyl acetate alcoholysis, catalyzed by immobilized CAL-B, was used for the synthesis of both isomers of calycotomine in a continuous process [11]. Asymmetric O-acylation of *N*-Boc-protected (6,7-dimethoxy-1,2,3,4-tetrahydroisoquinolin-1-yl)methanol was achieved, followed by separation of the enantiomers—the (*S*)-aminoalcohol and the (*R*)-amino ester. Next, deacylation and *N*-Boc-deprotection provided the desired (*S*)- and (*R*)-calycotomine, both in >99 % *ee*. Several amino alcohol homologues were also prepared (Scheme 9.2). Interestingly, increasing the distance between the hydroxyl group and the stereogenic center led to a drastic decrease of *E*—from 200 to 1 for a change of one to three carbon atoms.

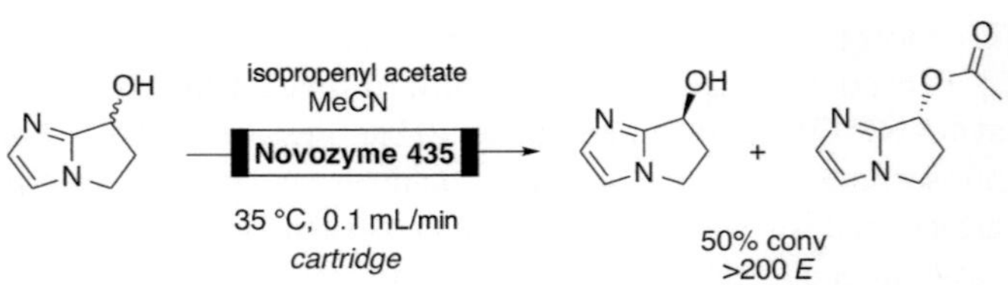

Scheme 9.1 Enzymatic kinetic resolution of 6,7-dihydro-5*H*-pyrrolo[1,2-a]imidazol-7-ol under flow conditions

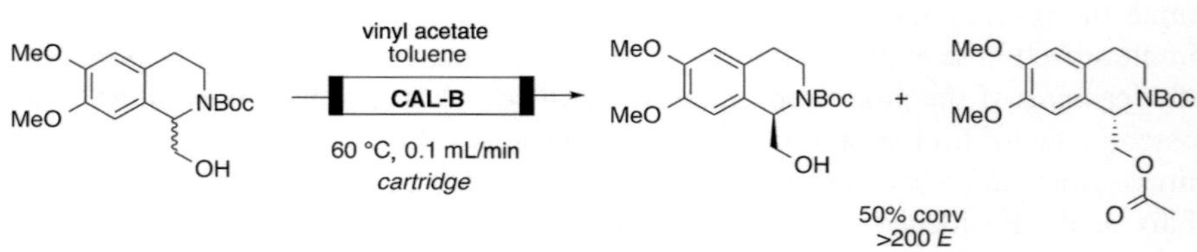

Scheme 9.2 Enzymatic kinetic resolution using immobilized CAL-B enzyme

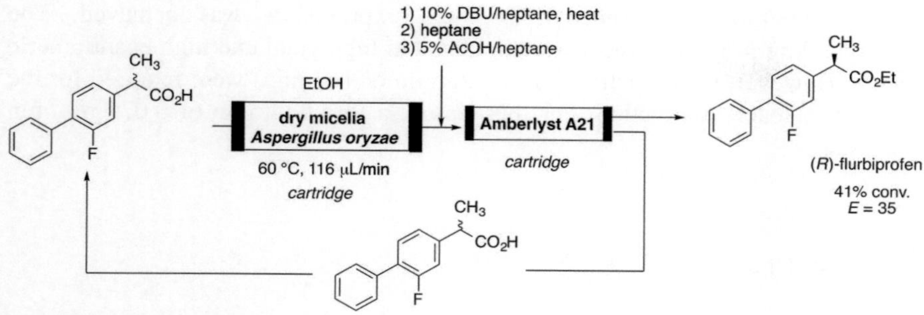

Scheme 9.3 A continuous flow kinetic resolution of (*RS*)-flurbiprofen

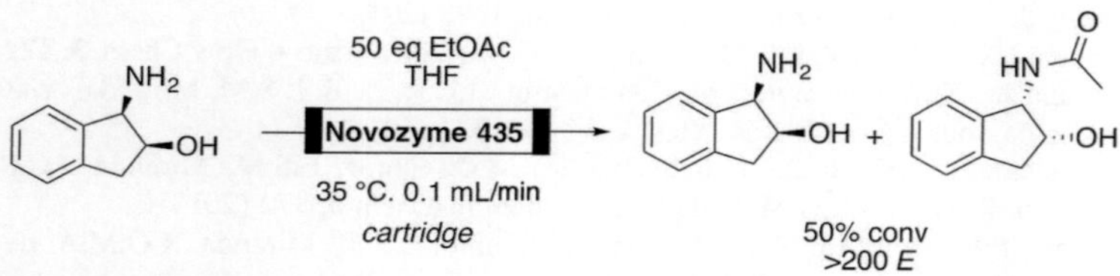

Scheme 9.4 Enzymatic kinetic resolution of a key precursor of indinavir

A whole microbial system—dry mycelium of *Aspergillus oryzae*—was used for the kinetic resolution of (*R*)-flurbiprofen via direct esterification with ethanol [12]. In a standard optimization procedure, the optimal reaction temperature (60 °C) and flow rate (116 μL/min) were determined. The final process implemented an in-line purification step by the catch-and-release principle. Thus, the (*R*)-flurbiprofen ethyl ester was easily obtained. The unreacted (*RS*)-flurbiprofen acid was simply recycled (Scheme 9.3).

The (1*S*,2*R*)-1-amino-2-indanol is a key intermediate for the synthesis of indinavir—an HIV protease inhibitor. Recently, an improved synthetic procedure was disclosed by Jeong et al. [13]. Using immobilized lipase B from *Candida antarctica*, enantiomerically pure *N*-acetyl-aminoindanol was conveniently prepared in a continuous flow procedure. A solution of (±)-*cis*-1-amino-2-indanol in EtOAc and THF (1:1) was fed to Novozyme 435 packed into a glass column and mounted on a flow reactor. After 64 min of residence time, the desired product was obtained with >99 % enantiomeric excess (Scheme 9.4).

9.2 Miscellaneous

An enantioselective bioreduction of β-ketoesters (ethyl 3-oxohexanoate and *tert*-butyl 3-oxobutanoate) using immobilized cells of *Kluyveromyces marxianus* or *Rhodotorula rubra* was carried out using a flow setup [14]. A significant reduction of reaction time as compared to batch experiments was achieved. The corresponding β-hydroxyesters were obtained in high yield and high enantiomeric excess. However, long reaction times (123 min or 164 min) were required for the reaction to reach completion. For this reason, a low flow rate of 0.075 mL/min was used.

References

1. C. Csajagi, G. Szatzker, E.R. Töke, L. Ürge, F. Darvas, L. Poppe, Tetrahedron Asymmetry **19**, 237 (2008)
2. I. Itabaiana Jr., M.C. Flores, F.K. Sutili, S.G.F. Leite, L.S. de M. Miranda, I.C.R. Leal, R.O.M.A. de Souza, Org. Process. Res. Dev. **16**, 1098 (2012)
3. I. Itabaiana Jr., I.C.R. Leal, L.S.M. Miranda, R.O.M.A. de Souza, J. Flow Chem. **3**, 122 (2013)
4. I. Itabaiana Jr., K.M. Goncalves, M. Zoumpanioti, I.C.R. Leal, L.S.M. Miranda, A. Xeankis, R.O.M.A. de Souza, Org. Process. Res. Dev. **18**, 1372 (2014)
5. K.M. Goncalves, F.K. Sutili, I. Itabaiana Jr., M.C. Flores, L.S.M. Miranda, I.C.R. Leal, Y. Cordeiro, R. Luque, R.O.M.A. de Souza, ChemSusChem **6**, 872 (2013)
6. H.S. Ruela, F.K. Sutili, I.C.R. Leal, N.M.F. Carvalho, L.S.M. Miranda, R.O.M.A. de Souza, Eur. J. Lipid Sci. Technol. **115**, 464 (2013)
7. I. Itabaiana Jr., F.K. Sutili, S.G.F. Leite, K.M. Goncalves, Y. Cordeiro, I.C.R. Leal, L.S.M. Miranda, M. Ojeda, R. Luque, R.O.M.A. de Souza, Green Chem. **15**, 518 (2013)
8. J. Madarasz, D. Nemeth, J. Bakos, L. Gubicza, P. Bakonyi, J. Clean. Prod. **93**, 140 (2015)
9. A.S. De Miranda, J.C. Gomes, M.T. Rodrigues Jr., I.C.R. Costa, W.P. Almeida, R. de O. Lopes, L.S.M. Miranda, F. Coelho, R.O.M.A. de Souza, J. Mol. Catal. B Enzym. **91**, 77 (2013)
10. E.A. Manoel, K.C. Pais, M.C. Flores, L.S.M. Miranda, M.A.Z. Coelho, A.B.C. Simas, D.M.G. Freire, R.O.M.A. de Souza, J. Mol. Catal. B Enzym. **87**, 139 (2013)
11. L. Schönstein, E. Forro, F. Fülöp, Tetrahedron Asymmetry **24**, 202 (2013)
12. L. Tamborini, D. Romano, A. Pinto, M. Contente, M.C. Iannuzzi, P. Conti, F. Molinari, Tetrahedron Lett. **54**, 6090 (2013)
13. Y.-J. Kim, Y.-S. Choi, S. Yang, W. Ram. Yang, J.-H. Jeong, Synlett **26**, 1981 (2015)
14. R.O. Lopes, S. Grimm, J.B. Ribeiro, I.C.R. Leal, L.S.M. Miranda, R.O.M.A. de Souza, J. Braz. Chem. Soc. **26**, 550 (2015)

Chapter 10
Organic Synthesis in Dedicated Continuous Flow Systems

Further Chemistry Examples

10.1 Photo- and Electrochemistry

Photochemistry is a valuable tool for the synthetic organic chemist, providing a unique path to otherwise not easily obtainable products. In recent years, the advances in both photochemistry and flow processing have led to increased amount of publications exploring the opportunities of a continuous photochemical process. Nevertheless, most of the published results were obtained in self-made or modified commercial devices (glass or quartz microreactors in combination with household light bulbs). Despite the fact that several dedicated instruments developed for photo-flow synthesis are available (e.g., Vapourtec, Future Chemistry), only few synthetic examples have been realized in such systems.

The [2 + 2]-photocycloaddition of a chiral cyclohexenone with cyclopentene was evaluated in a flow environment [1]. The photoreaction produces distorted cyclobutane skeletons that are not easily formed via the thermal pathway (Scheme 10.1). Using a commercial device equipped with a 1.5 W UV LED light source and able to precisely control the reaction temperature, the stereoselectivity of the reaction was intimately examined in detail. The flow reactor was superior to the batch equipment in terms of reaction efficiency even at high concentrations of the starting materials. To enhance the stereoselectivity, the addition of a 1-nitronaphthalene proved to be crucial.

Just recently, a flow electrochemistry cell targeted at synthetic laboratories was commercialized and a number of flow protocols were rapidly developed. The device was used to produce drug metabolites, thus simulating a CYP450-enzymatic oxidative transformation in the human liver [2]. Several commercial drug products were electrochemically "metabolized" on a 10–100 mg scale and the products were isolated and fully characterized.

Lately, the Shono oxidation of various N-protected cyclic amines was evaluated by Ley and coworkers [3]. The reaction proceeded best using a carbon anode and steel cathode in methanol at 0.1 M concentration, 43 mA current, and a flow rate of

T. Glasnov, *Continuous-Flow Chemistry in the Research Laboratory*,
DOI 10.1007/978-3-319-32196-7_10

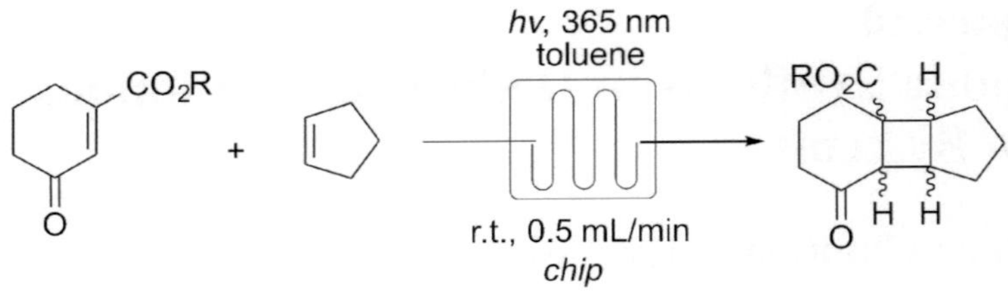

Scheme 10.1 A [2 + 2]-photocycloaddition of a chiral cyclohexenone with cyclopentene

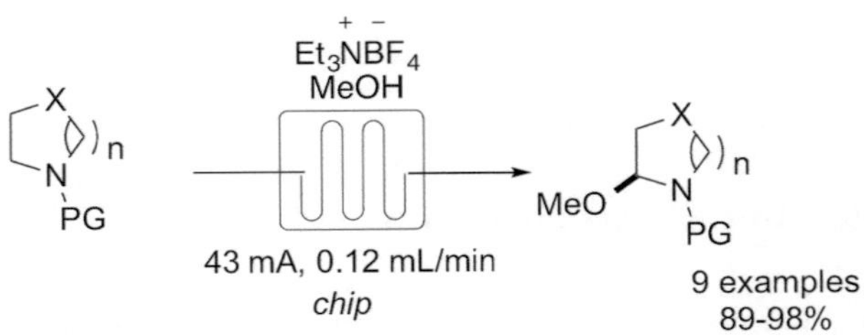

Scheme 10.2 Electrochemical Shono oxidation under flow conditions

120 μL/min with an isolated yield above 89 % (Scheme 10.2). The obtained α-methoxy amines were employed in a follow-up Pictet–Spengler reaction with tryptamines to provide several unnatural analogues of Nazinine—a tetrahydro-β-carboline with natural origin.

The Shono oxidation was used also to evaluate the influence of the volumetric flow rate, the variation of reactant concentration, charge output, rate of product formation, and possibilities to lower the electrolyte concentration under flow conditions in the flow electrochemistry cell [4].

10.2 Polymerization

The high susceptibility to reactor blockage is a major problem for continuous flow synthesis and is often associated with processing of highly viscous liquids such as polymer solutions. Nevertheless, numerous literature reports were committed on the evaluation of low-weight polymer synthesis under flow conditions.

Controlled radical polymerization using the reversible addition-fragmentation chain transfer approach (RAFT) was conducted in a continuous flow regime with series of different monomers—acrylamides, acrylates, and vinyl acetate [5]. The optimized conditions furnished 80–100 % conversions of the monomer at 70 °C or 100 °C reaction temperature. To avoid problems with developing overpressure, the reaction times had to be less than 2 h. Both batch and continuous flow processing delivered similar results.

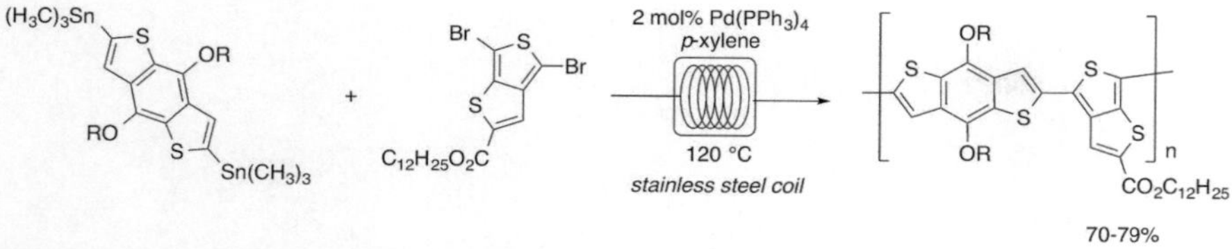

Scheme 10.3 Suzuki poly-condensation under flow conditions

Comparable results between batch and flow experiments were obtained also in the synthesis of conjugated polymers useful for organic photovoltaics prepared via a Suzuki or Kumada poly-condensation (Scheme 10.3) [5, 6].

In a very recent report, the synthesis of further polymers for organic photovoltaic applications was disclosed [7]. The high-performance benzodithiophene–thienopyrroledione copolymer PBDTTPD was synthesized via a Stille poly-condensation in chlorobenzene as a solvent. The synthesis provided 1.5 g of the desired product in a single flow run. The device-grade samples afforded a reproducible 7.2 % solar cell efficiency that could be increased up to 9.1 % upon addition of an ionic polythiophene-based cathodic interlayer.

Polymer endgroup modification and subsequent "click" conjugation were also achieved under flow conditions [8]. Well-defined block copolymers were easily accessed using azide- and alkyne-containing poly(butyl acrylate), poly(methyl acrylate), and polystyrene with full endgroup conversion. The "click" cycloaddition was carried out at 80 °C reaction temperature within 40 min reaction time.

Poly(2-oxazoline) triblock copolymers—otherwise difficult to access—were prepared via a continuous flow cascade [9]. Initially, the homopolymerization of 2-ethyl-2-oxazoline (EtOx) and 2-*n*-propyl-2-oxazoline (*n*PropOx) was optimized. A rapid reaction yielded well-controlled diblock copolymers at 160 °C in only 5 min reaction time. Varying the flow rate of the second monomer, different diblock copolymers could be obtained. Employing a second microreactor in a row, triblock copolymers of the structure EtOx-*b*-*n*PropOx-*b*-EtOx and *n*PropOx-*b*-EtOx-*b*-*n*PropOx were successfully prepared in a single process.

Using a similar reactor setup—a flow cascade of two microreactor chips—conjugated [2-methoxy-5-(3′,7′-dimethyloctyloxy)]-1,4-phenylenevinylene (MDMO-PPV) has been synthesized [10]. In a coupled process, the initial polymerization takes place at 50 °C, followed by an elimination step at 180 °C for 20 min. The obtained conjugated polymer showed a characteristic $\lambda_{max} = 500$ nm. A total monomer-to-conjugated MDMO-PPV conversion of 38 % was achieved.

10.3 Reactions Involving Organometallic Species

An inherently safe version of an amide bond formation using the pyrophoric trimethylaluminum has been reported in a flow system by Seeberger et al. [11]. Within a reaction time of only 2 min and 125 °C reaction temperature,

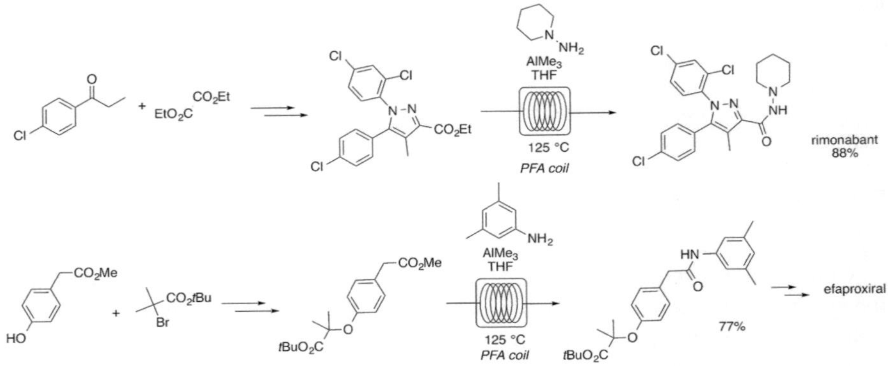

Scheme 10.4 Continuous flow access to rimonabant and efaproxiral

yields in the range of 37–98 % on a 0.2 mmol scale and with various substrates were achieved. A comparative study employing a batch microwave reactor delivered similar results. Furthermore, two sequential syntheses involving a trimethylaluminum-mediated amide bond formation were also disclosed, wherein the APIs rimonabant and efaproxiral were efficiently synthesized (Scheme 10.4).

Lately, the Bodroux amide formation involving a Grignard reagent was performed under continuous flow conditions as an alternative to already existing alkyl aluminum and other metal-catalyzed procedures [12]. With the help of i-PrMgCl-LiCl, the optimized reaction conducted at room temperature and for approximately 20 min total residence time enabled the generation of a small library of amides. A flow-based synthesis of sulfonamides was achieved by Gioiello et al. [13]. In a combinatorial fashion, five sulfonyl chlorides were reacted with 19 different amines in water/acetone/PEG400-solvent mixture, providing a library of sulfonamides with good yields. Based on traditional peptide synthesis methods, the flow coupling of activated building blocks was reported for the synthesis of some α-peptides via an amide bond formation [14]. The same group also demonstrated a continuous flow protocol for the preparation of amides based on the catalytic hydration of nitriles on the surface of a prepacked heterogeneous MnO_2 as a catalyst in the flow process [15]. The resulting amides were obtained by simply concentrating the output stream and without the need of further purification.

Organolithium chemistry is one of the most powerful and most versatile C–C bond forming reactions. A plethora of examples have already been performed in a continuous flow manner ("flash"-chemistry), however, only few in dedicated commercial flow reactors. The group of Schmalz elaborated an operationally simple and efficient lithium-mediated aryl carbinol synthesis starting from aryl bromides and ketones [16]. Using flow equipment, the coupling of 2-bromopyridine with fenchone was realized. At a temperature of −25 °C, the alcohol was obtained in 91 % yield. In contrast, cryogenic temperature of −78 °C was required while operating under batch conditions to achieve the same result. Thanks to the very efficient heat dissipation in the flow reactor, the halogen/lithium exchange could be

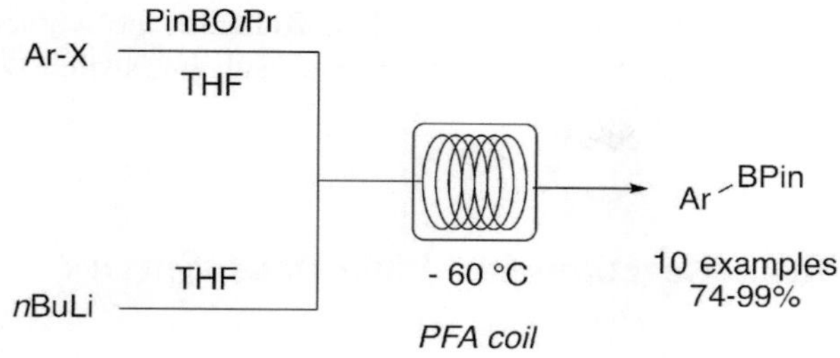

Scheme 10.5 Synthesis of boronic esters at cryogenic conditions

performed successfully even at ambient temperatures, which furthermore allowed reaction to be completed in less than a second and enabled high throughputs [17–19]. The synthesis of various boronic esters and acids via lithium exchange chemistry was utilized to evaluate a newly marketed reactor system, allowing comfortable synthesis at cryogenic conditions without the need of external cryostats (Scheme 10.5) [20, 21].

The concept was then extended to a broad range of low-temperature lithium exchange reactions [22]. In situ generated lithium diisopropylamide was used in a flow epimerization process to prepare all four stereoisomers of 3-(*tert*-butoxycarbonyl)-3-azabicyclo[3.1.0]hexane-2-carboxylic acid [23]. Regioselective lithiation–borylation and proton–lithium exchange at mild temperatures as and Knoevenagel condensation were successfully demonstrated by Wong et al. [24]. The flow metallation reaction of arenes and heteroarenes with the help of various metal salts—$ZnCl_2 \cdot 2LiCl$, $MgCl_2$, $CuCN \cdot 2LiCl$, and $LaCl_3 \cdot 2LiCl$—and in the presence of 2,2,6,6-tetramethylpiperidyl lithium was investigated by Knochel et al. and optimized under flow conditions to complete within 40 s residence time at 0 °C [25].

The Nobel Prize winning Grignard reaction, despite its age, is still a key approach to C–C bond formation. In the last decade, several synthetic examples of the Grignard addition were demonstrated under flow conditions in dedicated instrumentation. In a recent report by Rencurosi et al., the flow generation of Grignard reagents at room temperature followed by a coupling reaction with carbonyl compounds to prepare a small collection of secondary and tertiary alcohols has been devised [26]. The procedure was used also for the synthesis of the opioid analgesic Tramadol. The newly developed microscale ReactIR [27] flow cell was used as a convenient and versatile in-line spectroscopy analytical tool for monitoring the Grignard reaction in continuous flow environment. An intriguing halogen–magnesium exchange of four 3-iodoindoles followed by addition to three different aldehydes provided a collection of twelve 3-hydroxymethylindoles in a multistep flow setup [28]. The LiCl-mediated halogen–Mg exchange reaction was employed for the preparation of functionalized arylmagnesium compounds from aryl iodides or bromides via Grignard intermediates which were subsequently reacted with various carbonyl compounds [29]. Several perfluoroalkyl alkenes were synthesized via the telescoped Grignard reaction between (trimethylsilyl) methylmagnesium chloride and a trifluoromethyl ketone followed by dehydrative desilylation of the formed alcohol. To realize the synthetic sequence in continuous manner, an in-line extraction and solvent switch were implemented in the process [30].

10.4 Reactions Involving Diazo-Species

Diazo-compounds play an important role as reaction intermediates and have widespread use in synthetic organic chemistry. Nevertheless, the diazo-chemistry is generally considered hazardous due to the low stability and toxicity of the diazo-species. The growing number of continuous flow methods for the generation of various diazo-reagents allows to perform such chemistry under excellent control of reaction parameters in an inherently safe process environment.

A sequential flow process for the synthesis of β-ketoesters and subsequent condensation with amidines towards 2,6-substituted pyrimidine-4-ols was accomplished [31]. Using $BF_3{\cdot}OEt_2$ as a Lewis acid, high yields for the C–H insertion of ethyl diazoacetate into numerous aldehydes were achieved at ambient conditions. The crude β-ketoesters were then directly fed to a reaction with a number of amidines in the presence of an organic base to provide the corresponding 2,6-substituted pyrimidine-4-ols in 33–78 % overall yields (Scheme 10.6).

The hazards associated with diazo-compounds were also the motivation behind the studies of Hayes et al. [32]. Since only few diazo-compounds are commercially available, a sequential in situ synthesis from readily accessible aryl sulfonylhydrazones followed by immediate reaction with alcohols or amines to obtain α-amino or α-alkoxy acid derivatives has been developed (Scheme 10.7).

Various diazoesters were obtained under flow conditions at 80 °C or 99 °C in CH_2Cl_2 as a solvent, whereby quantitative conversions were easily achieved in almost all cases. The optimized protocol was then extended to include the subsequent O–H and N–H insertions, yielding the corresponding α-amino or α-alkoxy acid derivatives. The synthesis of α-sulfanyl, α-sulfonyl, and α-phosphono carboxylates has been accomplished via an in-line diazoesters formation from sulfonylhydrazones and a subsequent S–H, sulfinate, or P–H carbene insertion reactions [33]. Another difficult to handle diazo-reagent is diazomethane. Employing a base-induced decomposition reaction of *N*-methyl-*N*-nitrosourea in a continuous flow device, up to 19 mol/day of diazomethane could be generated [34]. The extension of the reaction system with an additional flow unit allowed a successful "capture" of the generated diazomethane with benzoic acid to provide the corresponding methyl ester. In a room temperature flow reaction and without the use of a transition metal catalyst, a number of aromatic hydrazones was transformed into the corresponding reactive diazo-compounds [35]. After passing a mixture of the hydrazone in the presence of an organic base over a packed column of activated MnO_2, the clean diazo-compound stream was combined with a stream

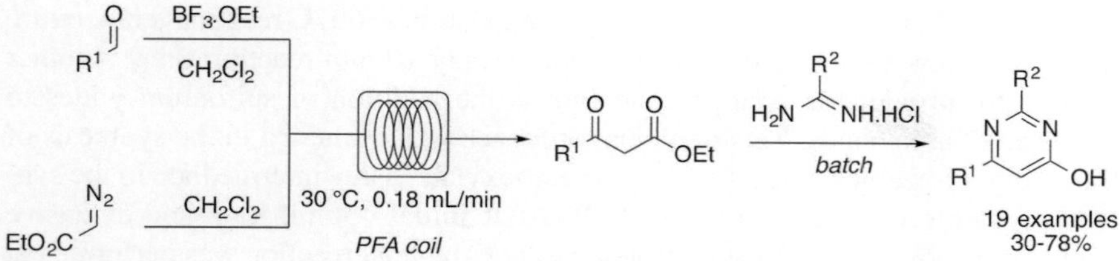

Scheme 10.6 Synthesis of 2,6-substituted pyrimidine-4-ols employing ethyl diazoacetate

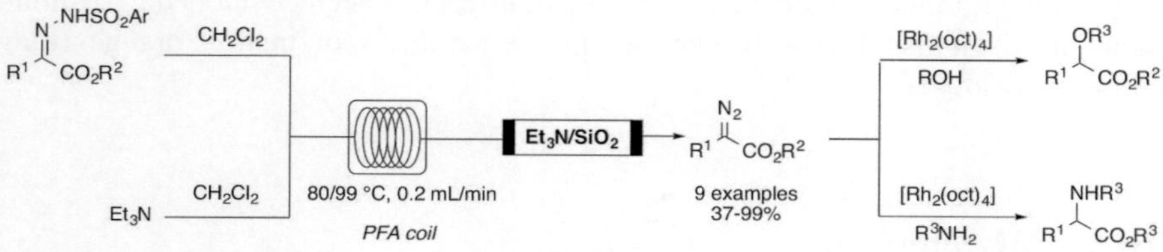

Scheme 10.7 Synthesis of α-amino or α-alkoxy acid derivatives via an in situ diazoester intermediate formation

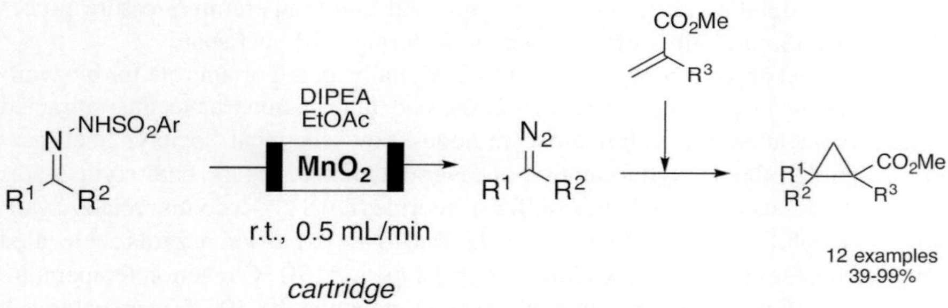

Scheme 10.8 Cyclopropanation of olefins wit diazo-compounds under flow conditions

of an olefin. Stirring at room temperature for 2 h provided the corresponding cyclopropane (Scheme 10.8).

The procedure was applied for the synthesis of a small library of cyclopropane ring containing esters and diols on a scale of up to 1.14 g. The same diazo-compound generation method was implemented in the preparation of allenes at ambient temperature [36]. Mixing the flow-generated unstable diazo-compounds with an alkyne in the presence of a copper catalyst enabled the smooth synthesis of numerous substituted allenes. A cyclopropane precursor useful in the synthesis of the antidepressant milnacipran was prepared in continuous flow via diazo-species using tosyl azide as reagent [37]. The process was previously investigated and optimized with the help of a FlowIR-unit in a number of N–H, O–H, and S–H insertions, as well as an N–H insertion of carbamates and a cyclopropanation by the same group [38]. An intramolecular cyclopropanation at 60 °C reaction temperature provided access to the desired product within only 10 min reaction time. Another method to introduce a cyclopropane ring is the addition of sulfonium ylides to activated double bonds. This particular approach was evaluated in the synthesis of trans-(dioxo)-azabicyclo-[3.1.0]-hexane carboxylate, a key intermediate in the synthesis of more complex molecules [39]. After initial optimization and extensive study of the possible side reactions and product, the final reaction was performed at 120 °C reaction temperature to provide 93 % yield.

Ultimately, a recent report underlined that flow equipment has finally found its way to undergraduate student labs—a flow protocol has been developed to demonstrate the benefits of flow technology in the synthesis of methyl orange from sulfanilic acid [40].

10.5 Miscellaneous

A broad range of classical batch chemistry and related synthetic procedures have been transformed into valuable continuous flow processes in the last decade—simple nucleophilic substitutions and high- and low-temperature/pressure procedures as well as a number of name reactions during a 10-year span.

Fluorinated organic compounds play traditionally an important role for bioavailability tuning in pharmaceutical synthesis, and fluorinations have thus attracted considerable attention within the continuous flow chemical society. Seeberger et al. explored the deoxyfluorination of several alcohols, sugars, carboxylic acids, and aldehydes using diethylaminosulfur trifluoride (DAST)—a commercially available nucleophilic fluorinating reagent [41]. Due to the explosion hazards at elevated temperatures, a safe flow procedure was developed. At 70 °C reaction temperature and in only 16 min residence time, yields in the range of 40–100 % were achieved. Identical results were disclosed by the Ley group, however, using a different flow system [42]. A library with 50 fluorinated compounds was synthesized using the developed method [43]. Instead of DAST, Selectfluor® or (1-chloromethyl-4-fluoro-1,4-diazoniabicyclo-[2.2.2]octane) bis(tetrafluoroborate) was used likewise an electrophilic reagent. A DAST-cyclodehydration procedure was utilized in the multistep flow synthesis of C_2-symmetric chiral PyBox ligands (Scheme 10.9) [44]. The synthesis starts with commercial chelidonic acid which is transformed

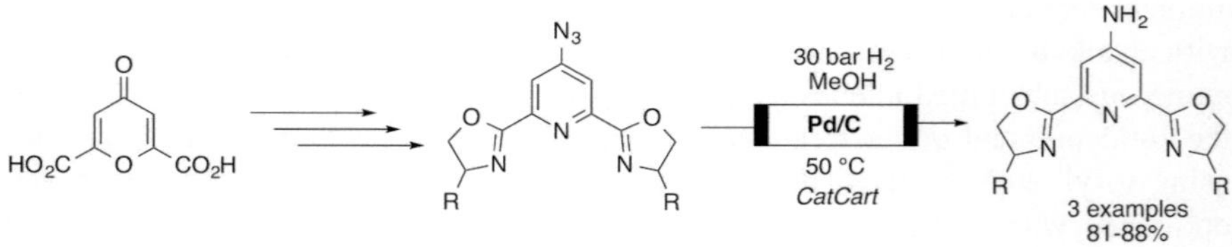

Scheme 10.9 Continuous flow multistep synthesis of C_2-symmetric PyBox ligands

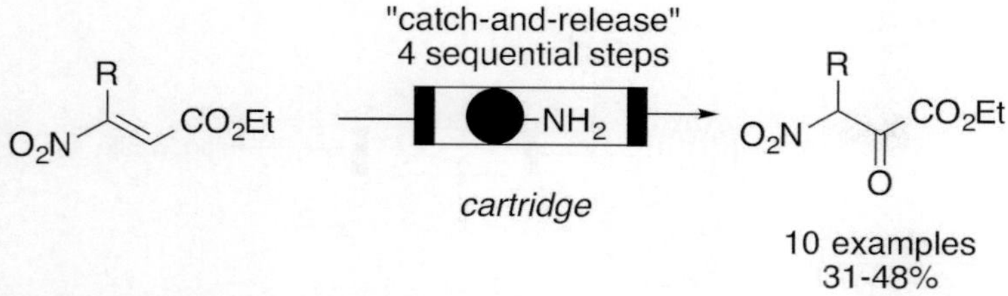

Scheme 10.10 Synthesis of useful α-ketoester under flow conditions

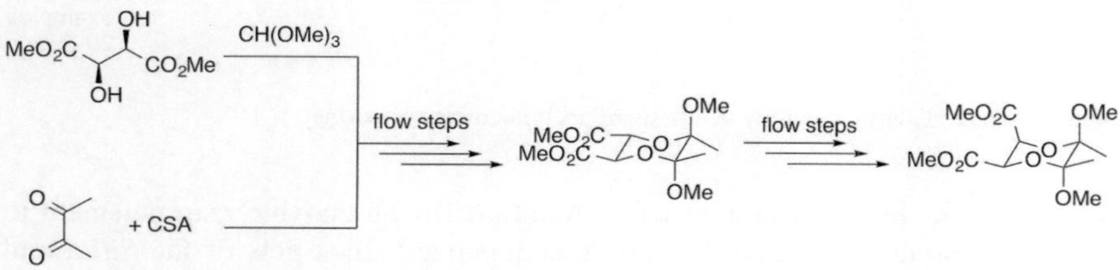

Scheme 10.11 Example of a flow butane-2,3-diacetal protection of an α-hydroxy acid

into chelidamic acid using 25 % aqueous ammonia, followed by Vilsmeier chlorination, bisamidation, cyclodehydration, azidation, and reduction steps, all performed under flow conditions.

A phase-transfer alkylation of phenylacetonitrile was used to evaluate the performance of the Ehrfeld microreactor platform in direct comparison with a batch process [45].

The phase-transfer synthesis of O- and S-benzyl ethers served the same purpose in a study to assess the performance of different microreactor types in a comparison with a microwave batch reactor [46]. In addition, the same group evaluated the phase-transfer chlorodehydroxylation in an uncatalyzed high-temperature regime [47]. The alkylation of an extended library of thiols was reported also by Wirth et al. [48]. Another phase-transfer process resulted in the successful dibromocyclopropanation of a representative selection of alkenes with different structural features [49]. The N-alkylation of aromatic amines was most recently evaluated in analogous fashion most recently [50]. A slightly improved ratio of mono- to di-alkylated products was observed in comparison to the batch process.

Starting from a range of nitroalkanes, synthetically useful α-ketoester adducts were synthesized in a Henry reaction under flow conditions (Scheme 10.10) [51].

The same group reported the synthesis of butane-2,3-diacetal protected building blocks using flow processing with implemented IR-spectroscopic analysis and employing supported reagents and scavengers (Scheme 10.11) [52, 53].

The reverse process—diacetal deprotection—has been performed with a novel heterogeneous catalyst—a mesoporous silica-supported Er(III) [54]. The solid catalyst was prepared under microwave irradiation and packed into CatCarts®

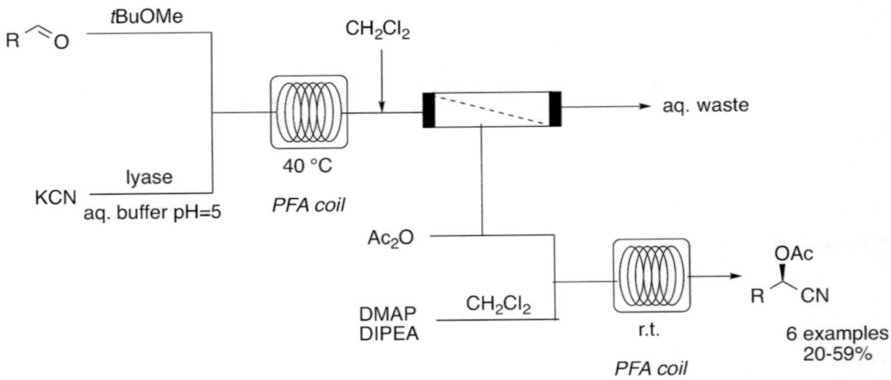

Scheme 10.12 Continuous flow synthesis of acetylated cyanohydrins

prior its evaluation in a flow process. Additionally, an epoxide rearrangement to aldehyde was demonstrated. Various functionalized silica gels of the *Silia*Bond series such as acid, base, and nucleophilic catalysts have been likewise examined in several flow examples in the form of solid-state reagents [55].

Hydroxamic acids occur in several molecules displaying a spectrum of biological activities and are thus of potential interest for the pharmaceutical industry. A general flow procedure for the conversion of esters, hydroxylamine, and sodium methoxide into substituted hydroxamic acids in good yields and purities was developed [56].

In a cooperation between industry and academia, the three-step batchwise synthesis of the pharmaceutical intermediate (1*R*,2*S*,4*S*)-(7-oxa-bicyclo[2.2.1]hept-2-yl)-carbamic acid ethyl ester was translated to a continuous flow process comprising a highly exothermic hydrazine quenching step and an acylazide generation step which were managed safely and delivered high yields [57].

The synthesis of the N-Boc-protected 3,4-dehydro-L-proline methyl ester was accomplished with the help of a modular flow reactor in 87 % overall yield, 97 % purity, and >98 % enantiomeric excess. Immobilized reagents and scavengers were used to produce 9 g of the target compound in 12 h [58]. Fmoc protection and debenzylation were exploited for the synthesis of N-Fmoc-(6-Boc-aminohexyl)glycine and N-Fmoc-((2-(2-Bocaminoethoxy)ethoxy)ethyl)glycine in a flow environment [59].

Acetone cyanohydrin is a widely used liquid source of hydrogen cyanide. Its synthesis requires a safe protocol to handle the highly toxic in situ generated HCN.

A simple, benchtop flow system has been employed by Stevens et al. to achieve an output of 39 g/h acetone cyanohydrin with quantitative conversion of acetone in the reaction with KCN and acetic acid [60]. A slightly modified process was used by Rutjes et al. to prepare acetylated cyanohydrins (Scheme 10.12) [61].

Bronsted acid catalyzed glycosylation for the synthesis of nucleosides was demonstrated by the group of Jamison. A glycosyl donor with a silylated

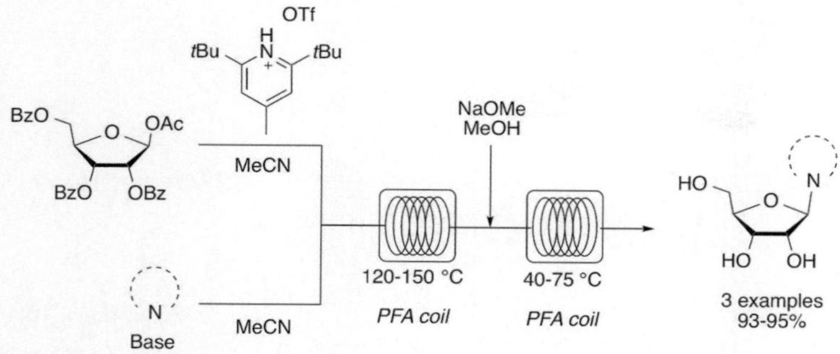

Scheme 10.13 Continuous flow synthesis of nucleosides

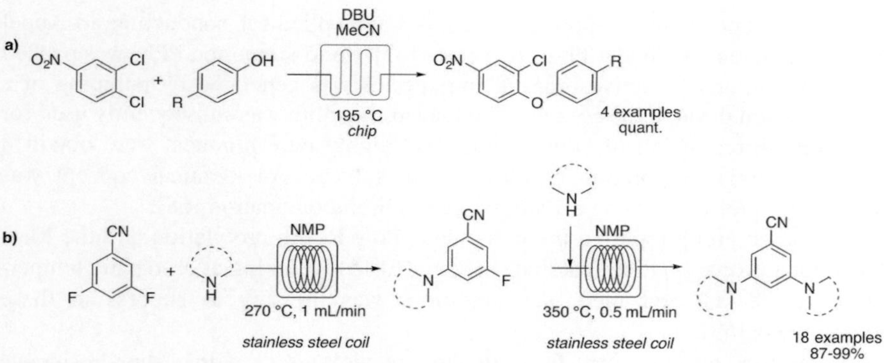

Scheme 10.14 Examples of S_NAr substitution reactions under flow conditions

nucleobase was assembled in either a batch microwave or a flow experiments in comparable yields (Scheme 10.13) [62].

The high-temperature S_NAr diaryl ether synthesis from aryl chlorides and phenols was demonstrated as direct translation of a batch microwave method into a flow process (Scheme 10.14a) [63]. In a similar fashion and using another benchtop reactor, 3,5-diamino-benzonitriles were prepared from 3,5-difluoro-benzonitriles and various amines. Temperatures above 250 °C were required to achieve full conversions for all examples of the small library generated (Scheme 10.14b) [64]. The same reaction was used to demonstrate the coupling of a flow system with a simulated moving bed chromatography [65].

A further example of S_NAr substitution is the straightforward ligand-free synthesis of phenols from the corresponding aryl iodides in a flow system [66]. The reactions requires only 4–20 min at 150–165 °C reaction temperature to reach competition, furnishing 40–87 % yields of the corresponding products. Instead of using a copper catalyst, a commercially available copper coil was utilized as reactor and catalyst source at once.

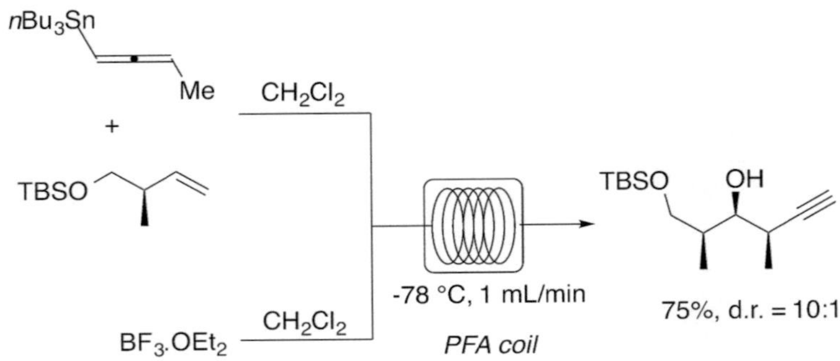

Scheme 10.15 An example of a Marshall homopropargylation reaction under flow conditions

The concept of solid-supported reagents was applied for conducting an Appel bromination reaction under flow conditions [67]. Solid-supported PPh_3 was packed in a column, and an active brominating species was generated by pumping of a CBr_4 solution through the column. The loaded monolith was subsequently used for the Appel bromination of various alcohols. Highly pure products were obtained through simple evaporation of the reaction solvent. An identical concept was applied also for conducting a Ramirez *gem*-dibromoolefination [68].

The same group reported the continuous flow Roush crotylation and the Marshall homopropargylation reaction (Scheme 10.15). Working at cryogenic temperatures (−78 °C) and inert gas conditions was the key to success in these experiments [69].

To prepare nitrous esters from alcohols in high yields within short residence time, an efficient mass transfer and tight control of internal reaction temperatures are required to master highly exothermic conditions. Reactions of this kind were successfully conducted by Monbaliu et al. in a Corning® Advanced-Flow™ reactor [70]. At a reaction temperature of 18 °C, the controlled mixing of an alcohol, HCl, and aq. $NaNO_2$ provided the desired nitrous ester products. Nitration of aromatic compounds with fuming HNO_3 is another challenge. Performing such reactions under flow conditions requires special equipment to not only dissipate the released heat from the process but also to prevent reactor fouling. A research group at Novartis faced the challenge and demonstrated a safe continuous process for the nitration of various important intermediates on a gram scale [71]. A nitration employing fuming HNO_3 was also reported by Yu et al. for the synthesis of 2,5-difluoronitrobenzene [72].

Continuous flow processing has been of rising interest in the field of organocatalysis as well. A packed-bed of L-proline was used as a slowly leaching catalyst source [73] to demonstrate continuous proline catalyzed α-aminoxylations. The reactions were accomplished at a temperature of 0–5 °C and reached high conversions with enantiomeric excess of nearly 98 % *ee*-value. The selective asymmetric synthesis of γ-nitroaldehydes, utilizing solid-supported tripeptide catalysts, was reported by Fülöp et al. [74]. The catalysts were readily synthesized, immobilized, packed in CatCarts®, and tested in the 1,4-addition of propanal to

E-β-nitrostyrene. While requiring up to 24 h of reaction time, batch experiments provided better conversions along with similar enantioselectivity in comparison with the flow experiments which had a reaction time of only 7 min. A solid-supported cinchona-squaramide bifunctional catalyst was prepared and utilized instead of peptides for a similar organocatalytic reaction—the conjugate addition of diketones and nitrostyrenes [75]. Yields above 75 % and *ee*-values above 96 % were achieved. An identical reaction set was evaluated for the synthesis of a precursor of the $GABA_B$ receptor agonist baclofen [76].

Radical-based reactions are another type of valuable synthetic transformations. Due to the hazardous properties of many of the typically utilized reagents, continuous flow processes have been evaluated. A highly efficient system for tris (trimethylsilyl)silane (TTMSS)-mediated Barton–McCombie deoxygenation, dehalogenation, and hydrosilylation reactions has been established (Scheme 10.16a) [77]. TTMSS was used as a nontoxic alternative to tin hydrides traditionally employed as the reducing agent and delivered yields above 67 %. In a similar study, radical dehalogenation or dehalogenation–cyclization reactions were explored under flow conditions using TTMSS [78]. A gram-scale synthesis of a natural product precursor was demonstrated wherein a total amount of 7.6 g product

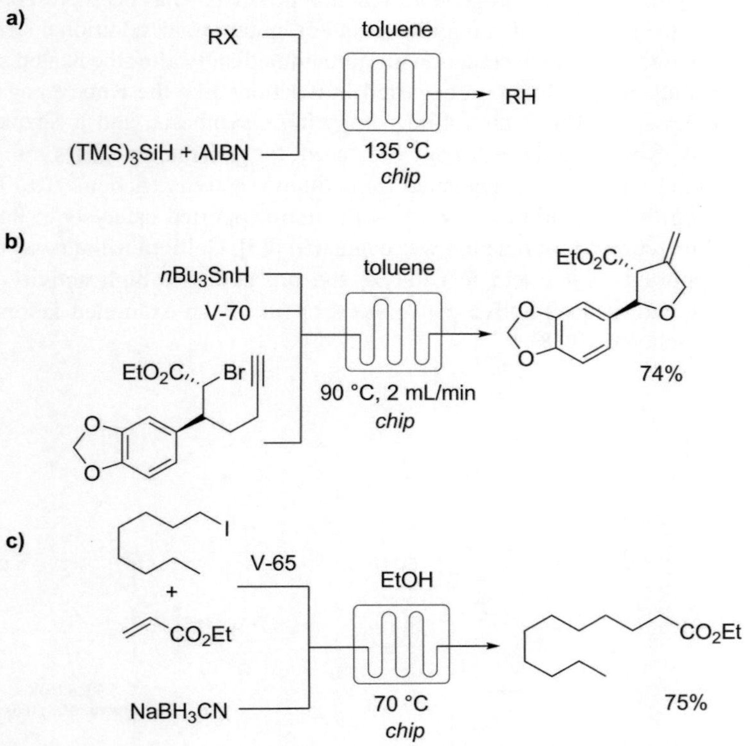

Scheme 10.16 Various radical processes under flow conditions

(74 %) could be isolated after 185 min and chromatography purification (Scheme 10.16b). The cyanoborohydride-mediated Giese reaction of several alkyl iodides with ethyl acrylate was studied in a continuous flow system by the same group [79]. Using the optimized flow conditions (70 °C, 10–15 min reaction time), high yields of the corresponding adducts could be obtained (Scheme 10.16c).

Transforming biological feed stocks into useful chemical intermediates or fuels grows of increasing importance. As many of the relevant sustainable or green chemistry examples are high-temperature processes, this area of research has found new opportunities in the use of flow processing. A highly efficient methodology for the dehydration of sucrose, D-fructose, and D-glucose to the furan derivatives such as 5-(chloromethyl)furfural (CMF) and 5-(hydroxymethyl)furfural (HMF) as well as levulinic acid (LA) by continuous flow processing was recently reported [80]. Using the new technology, high yields (>60 %) in short reaction times (1–15 min) were achieved. An identical study by Mihovilivic et al. confirmed the obtained results [81]. A chemocatalytic process for the valorization of glyoxal, an undesired oxygenated component of pyrolysis oil, was investigated under flow conditions [82]. The aqueous-phase isomerization of glyoxal into glycolic acid was catalyzed by gallium- or tin-doped zeolites as heterogeneous catalysts.

An interesting solution to a common problem in flow synthesis—reactor blockage by precipitation of a heterogeneous reaction products—has been presented in a recent study [83]. Instead of using ultrasound or agitation, an additional stream of solvent was used to dilute the reaction mixture immediately after the heated zone of the rector; thus, no precipitation occurred in reactions like the Knoevenagel coumarin synthesis, the Hantzsch 1,4-dihydropyridine synthesis, and a Suzuki C–C coupling. A Knoevenagel reaction was used to generate a series of (20*R*)-panaxadiol derivatives to be evaluated for antitumor activity (Scheme 10.17) [84].

Exploiting the possibilities to work with solid-supported catalysts in flow, the Strecker three-component reaction was evaluated [85]. Gallium triflate was used as a solid-supported Lewis acid to catalyze the process. The high activity of the catalyst resulted in quantitative conversions of all of the examined ketones and aldehydes (Scheme 10.18).

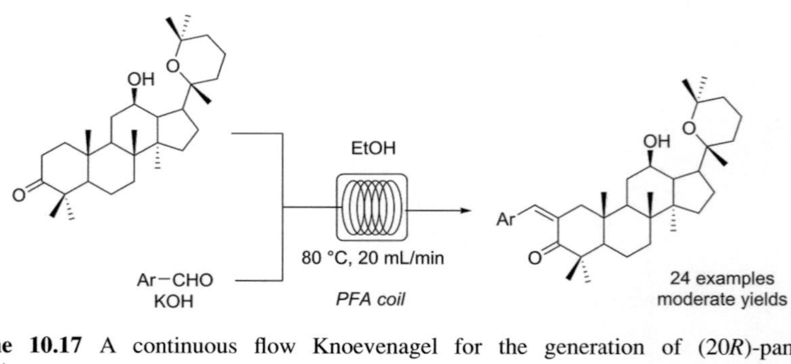

Scheme 10.17 A continuous flow Knoevenagel for the generation of (20*R*)-panaxadiol derivatives

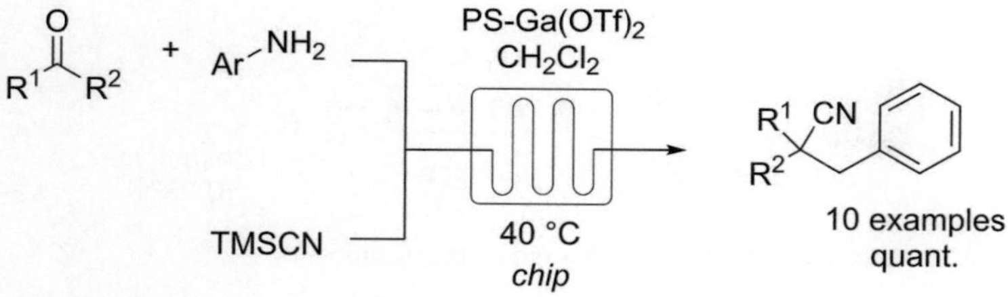

Scheme 10.18 Strecker three-component reaction under flow conditions

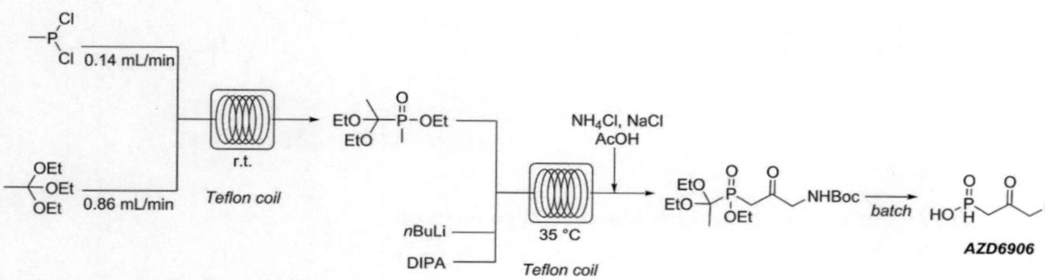

Scheme 10.19 Stepwise flow/batch approach towards AZD6906

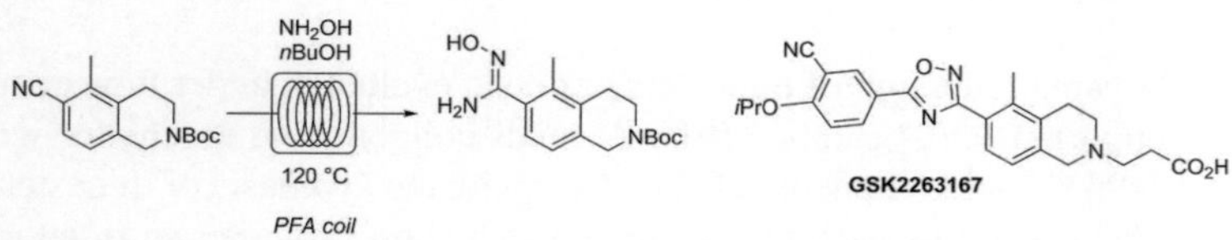

Scheme 10.20 Flow synthesis of a synthetic intermediate of GSK2263167

A promising reflux inhibitor—AZD6906—was synthesized in a combined flow and batch procedure for the evaluation of scalable synthetic routes [86]. A robust method to handle toxic/reactive reagents was developed, allowing the process to be performed at 35 °C and in very short reaction times—1–2 min (Scheme 10.19).

The Vilsmeier–Haack formylation has been performed as a flow process to prepare a few formylated products in low to moderate yields [87]. Optimal results were obtained at 60 °C reaction temperature and reaction time of 3 min.

Aromatic amidoximes were prepared safely in a flow instrument from the reaction of the corresponding nitriles with hydroxylamine. With the help of a design-of-experiment (DoE) software, the reaction conditions were meticulously optimized [88]. An identical approach was implemented in the synthesis of the $S1P_1$ receptor agonist GSK2263167 (Scheme 10.20) [89].

Readily available chloroximes were used as the starting materials for the synthesis of various isothiocyanates in flow [90]. The desired isothiocyanate products were obtained via a 1,3-dipolar cycloaddition reaction between a nitrile oxide and a thiourea compound, 1,4,2-oxathiazoline rearrangement, and urea elimination. Two immobilized reagents (a weak base and a functionalized thiourea) were loaded as a

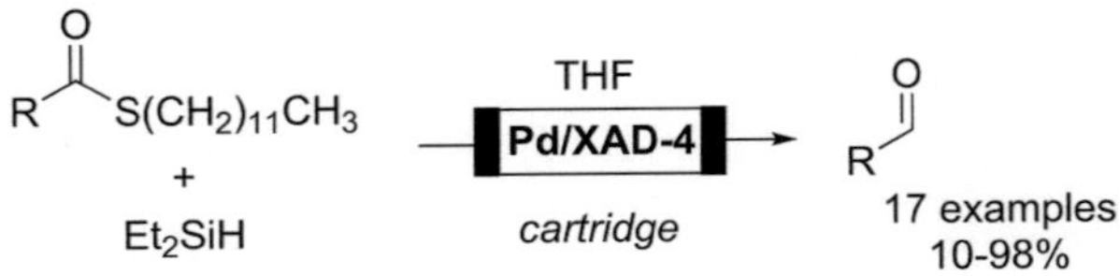

Scheme 10.21 Continuous flow Fukuyama reduction of thioesters

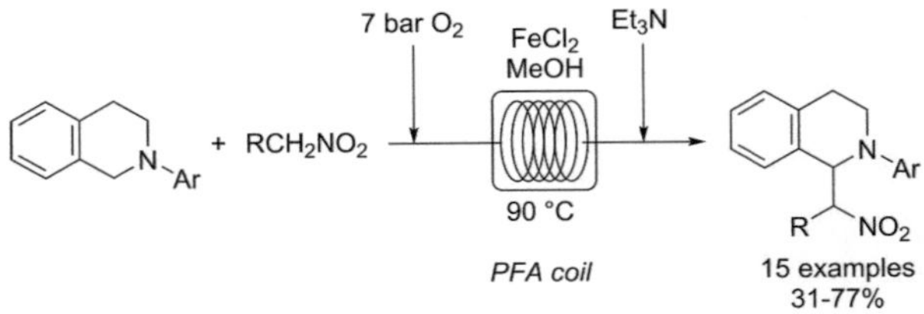

Scheme 10.22 Continuous flow nitro-Mannich type reaction

1:1-mixture into a single glass cartridge to promote the synthesis. Using supported reagents allowed the simple isolation of the product by simple solvent evaporation. The room temperature process provided the desired isothiocyanates in 71–93 % yields.

The Fukuyama reduction of thioesters has been evaluated under flow conditions utilizing supported Pd(0) catalyst [91]. A small collection of aldehydes was prepared in 10–96 % yields (Scheme 10.21). Combing the synthesis of thioesters from acyl chlorides and thiols with the optimized reduction provided an uninterrupted synthesis of aldehydes from the corresponding acid chlorides.

Employing various supported reagents, a small library of secondary and tertiary sulfonamides with expected selective carbonic anhydrase IX and XII inhibitory activity was easily generated from sulfonyl chlorides and the corresponding amines operating a flow system at room temperature [92]. The somewhat challenging amide formation starting from low-nucleophilic amines and esters was successfully optimized in a flow experiment [93]. Using a strong base—LiHMDS—and the efficient mixing promoted the reaction to achieve yields of in the range of 46–100 %. In addition, an extended procedure allowed the flow generation of the corresponding esters, followed by the subsequent formation of amides in an uninterrupted flow process. Amides and esters were generated via an efficient generation and trapping of ketenes under flow conditions [94]. Employing EtO-, *i*PrO-, and *t*BuO alkoxyalkynes, a thermolysis at 140–180 °C led to the generation of the corresponding ketenes, which were then immediately reacted with alcohols or amines to deliver the final products in high yields.

An efficient nitro-Mannich type direct α-C(sp^3)–H functionalization of N-aryl-1,2,3,4-tetrahydroisoquinolines catalyzed by simple iron salts and using molecular oxygen as the terminal oxidant and in a flow environment (Scheme 10.22) [95].

In order to introduce molecular oxygen into the reaction mixture, a Teflon AF-2400 membrane tube-in-tube reactor was utilized for reaction intensification.

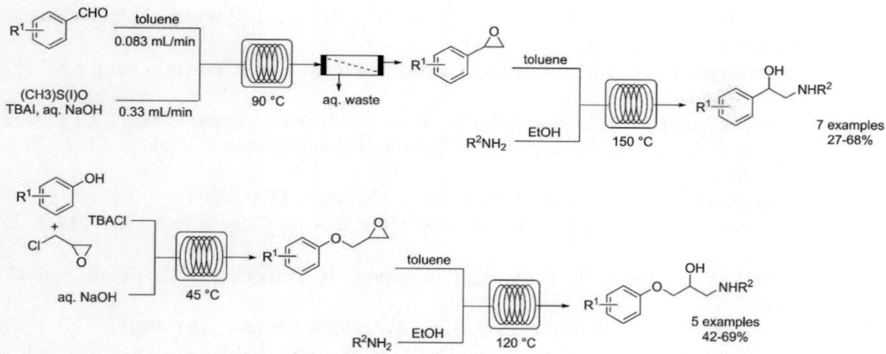

Scheme 10.23 Continuous flow generation and ring opening of epoxides with amines

At 90 °C reaction temperature and an oxygen pressure of seven bar, the reaction provided a set of N-aryl-tetrahydroisoquinoline nitro-Mannich adducts.

The generation and ring opening of epoxides was exploited for the continuous flow access to vicinyl amino alcohols (Scheme 10.23) [96]. The epoxidation of benzaldehyde was utilized for the optimization of the Corey–Chaykovsky flow reaction. The subsequent aminolysis of the formed epoxide ring provided the bronchodilator tulobuterol in 27 % overall yield. Using epichlorohydrin as reagent, an epoxide ring was installed on several selected alcohols. These were subsequently subjected to the aminolysis conditions to produce further examples of active pharmaceutical ingredients—propranolol, metoprolol, alprenolol, and bupranolol.

The synthesis of (*E*)-(*S*)-3-hydroxy-7-tritylthio-4-heptenoic acid, a key component of cyclodepsipeptide histone deacetylase inhibitors, was performed using a flow setup [97]. The 1,4-addition of trityl thiol to acrolein, a Horner–Wadsworth–Emmons reaction, a partial reduction, and a diastereoselective aldol reaction in a sequence were investigated under flow conditions.

Several reactions have been used to exemplify technical modifications towards further automation of continuous flow synthesis, based on available commercial devices [98–100].

References

1. K. Terao, Y. Nishiyama, S. Aida, H. Tanimoto, T. Morimoto, K. Kakiuchi, J. Photochem. Photobiol. A Chem. **242**, 13 (2012)
2. R. Stalder, G.P. Roth, ACS Med. Chem. Lett. **4**, 1119 (2013)
3. M.A. Kabeshov, B. Musio, P.R.D. Murray, D.L. Browne, S.V. Ley, Org. Lett. **16**, 4618 (2014)
4. R.A. Green, R.C.D. Brown, D. Pletcher, J. Flow Chem. **5**, 31 (2015)
5. H. Seyler, D.J. Jones, A.B. Holmes, W.W.H. Wong, Chem. Commun. **48**, 1598 (2012)
6. H. Seyler, J. Subbiah, D.J. Jones, A.B. Holmes, W.W.H. Wong, Beilstein J. Org. Chem. **9**, 1492 (2013)

7. G. Pirotte, J. Kesters, P. Verstappen, S. Govaerts, J. Manca, L. Lutsen, D. Vanderzande, W. Maes, ChemSusChem **8**, 3228 (2015)
8. J. Vandenbergh, T. Tura, E. Baeten, T. Junkers, J. Polym. Sci. A Polym. Chem. **52**, 1263 (2014)
9. E. Baeten, B. Verbraeken, R. Hoogenboom, T. Junkers, Chem. Commun. **51**, 11701 (2015)
10. N. Zaquem, E. Baeten, J. Vandenbergh, L. Lutsen, D. Vanderzande, T. Junkers, Chem. Eng. Technol. **38**, 1749 (2015)
11. T. Gustafsson, F. Ponten, P.H. Seeberger, Chem. Commun. 1100 (2008)
12. J. M. Munoz, J. Alcazar, A. de la Hoz, A. Diaz-Ortiz, S-A. A. Diego, Green Chem. **14**, 1335 (2012)
13. A. Gioiello, E. Rosatelli, M. Teofrasti, P. Filipponi, R. Pellicciari, ACS Comb. Sci. **15**, 235 (2013)
14. K.R. Knudsen, M. Ladlow, Z. Bandpey, S.V. Ley, J. Flow Chem. **4**, 18 (2014)
15. C. Battilocchio, J.M. Hawkins, S.V. Ley, Org. Lett. **16**, 1060 (2014)
16. S. Goto, J. Velder, S.E. Sheikh, Y. Sakamoto, M. Mitani, S. Elmas, A. Adler, A. Becker, J-M. Neudoerfl, J. Lex, H.-G. Schmalz, Synlett 1361 (2008)
17. D.W. Slocum, D. Kusmic, J.C. Raber, T.K. Reinscheld, P.E. Whitley, Tetrahedron Lett. **51**, 4793 (2010)
18. D.W. Slocum, K.C. Tekin, Q. Nguyen, P.E. Whitley, T.K. Reinscheld, B. Fouzia, Tetrahedron Lett. **52**, 7141 (2011)
19. D.W. Slocum, K. Reinscheld, N.D. Austin, D. Kusmic, P.E. Whitley, Synthesis **44**, 2531 (2012)
20. D.L. Browne, M. Baumann, B.H. Harji, I.R. Baxendale, S.V. Ley, Org. Lett. **13**, 3312 (2011)
21. J.A. Newby, L. Huck, D.W. Blaylock, P.M. Witt, S.V. Ley, D.L. Browne, Chem. Eur. J. **20**, 263 (2014)
22. J.A. Newby, D.W. Blaylock, P.M. Witt, J.C. Pastre, M.K. Zacharova, S.V. Ley, D.L. Browne, Org. Process Res. Dev. **18**, 1211 (2014)
23. B. Bakonyi, M. Furegati, C. Kramer, L. La Vecchia, F. Ossola, J. Org. Chem. **78**, 9328 (2013)
24. H. Seyler, S. Haid, T.-H. Kwon, D.J. Jones, P. Baeuerle, A.B. Holmes, W.W.H. Wong, Aust. J. Chem. **66**, 151 (2013)
25. M.R. Becker, P. Knochel, Angew. Chem. Int. Ed. **54**, 12501 (2015)
26. E. Riva, S. Gagliardi, M. Martinelli, D. Passarella, D. Vigo, A. Rencurosi, Tetrahedron **66**, 3242 (2010)
27. C.F. Carter, H. Lange, S.V. Ley, I.R. Baxendale, B. Wittkamp, J.G. Goode, N.L. Gaunt, Org. Process Res. Dev. **14**, 393 (2010)
28. T. Tricotet, D.F. O'Shea, Chem. Eur. J. **16**, 6678 (2010)
29. T. Brodmann, P. Koos, A. Metzger, P. Knochel, S.V. Ley, Org. Process Res. Dev. **16**, 1102 (2012)
30. T.A. Hamlin, G.M.L. Lazarus, C.B. Kelly, N.E. Leadbeater, Org. Process Res. Dev. **18**, 1253 (2014)
31. H.E. Bartrum, D.C. Blakemore, C.J. Moody, C.J. Hayes, J. Org. Chem. **75**, 8674 (2010)
32. H.E. Bartrum, D.C. Blakemore, C.J. Moody, C.J. Hayes, Chem. Eur. J. **17**, 9586 (2011)
33. H.E. Bartrum, D.C. Blakemore, C.J. Moodya, C.J. Hayes, Tetrahedron **69**, 2276 (2013)
34. E. Rossi, P. Woehl, M. Maggini, Org. Process Res. Dev. **16**, 1146 (2012)
35. N.M. Roda, D.N. Tran, C. Battilocchio, R. Labes, R.J. Ingham, J.M. Hawkins, S.V. Ley, Org. Biomol. Chem. **13**, 2550 (2015)
36. J.-S. Poh, D.N. Tran, C. Battilocchio, J.M. Hawkins, S.V. Ley, Angew. Chem. Int. Ed. **54**, 7920 (2015)
37. S.T.R. Müller, A. Murat, P. Hellier, T. Wirth, doi: 10.1021/acs.oprd.5b00308
38. S.T.R. Mìller, A. Murat, D. Maillos, P. Lesimple, P. Hellier, T. Wirth, Chem. Eur. J. **21**, 7016 (2015)
39. F.G. Buono, M.C. Eriksson, B.-S. Yang, S.R. Kapadia, H. Lee, J. Brazzillo, J.C. Lorenz, L. Nummy, C.A. Busacca, N. Yee, C. Senanayake, Org. Process Res. Dev. **18**, 1527 (2014)

40. L. van Rens, H. van Dijk, J. Mulder, P. Nieuwland, J. Chem. Educ. **90**, 574 (2013)
41. T. Gustafsson, R. Gilmour, P.H. Seeberger, Chem. Commun. 3022 (2008)
42. M. Baumann, I.R. Baxendale, S.V. Ley, Synlett 2111 (2008)
43. M. Baumann, I.R. Baxendale, L.J. Martin, S.V. Ley, Tetrahedron **65**, 6611 (2009)
44. C. Battilocchio, M. Baumann, I.R. Baxendale, M. Biava, M.O. Kitching, S.V. Ley, R.E. Martin, S.A. Ohnmacht, N.D.C. Tappin, Synthesis **44**, 635 (2012)
45. E.S. Borovinskaya, L. Mammitzsch, V.M. Uvarov, F. Schael, W. Reschetilowski, Chem. Eng. Technol. **32**, 919 (2009)
46. B. Reichart, C.O. Kappe, T.N. Glasnov, Synlett 2393 (2013)
47. B. Reichart, G. Tekautz, C.O. Kappe, Org. Process Res. Dev. **17**, 152 (2013)
48. A. Baker, M. Graz, R. Saunders, G.J.S. Evans, I. Pitotti, T. Wirth, J. Flow Chem. **5**, 65 (2015)
49. R.B. Østby, Y.H. Stenstrøm, T. Didriksen, J. Flow Chem. **5**, 69 (2015)
50. Y.-S. Choia, Yoon-Jung Kima, L.-L. Shena, Y. S. Lee, J.-H. Jeong, Synlett 970 (2015)
51. A. Palmieri, S.V. Ley, A. Polyzos, M. Ladlow, I.R. Baxendale, Beilstein J. Org. Chem. **5**, 23 (2009). doi:10.3762/bjoc.5.23
52. C.F. Carter, I.R. Baxendale, M. O'Brien, J.B.J. Pavey, S.V. Ley, Org. Biomol. Chem. **7**, 4594 (2009)
53. C.F. Carter, I.R. Baxendale, J.B.J. Pavey, S.V. Ley, Org. Biomol. Chem. **8**, 1588 (2010)
54. M. Oliverio, A. Procopio, T.N. Glasnov, W. Goessler, C.O. Kappe, Aust. J. Chem. **64**, 1522 (2011)
55. V. Pandarus, G. Gingras, F. Beland, R. Ciriminna, M. Pagliaro, Catal. Sci. Technol. **1**, 1600 (2011)
56. E. Riva, S. Gagliardi, C. Mazzoni, D. Passarella, A. Rencurosi, D. Vigo, M. Martinelli, J. Org. Chem. **74**, 3540 (2009)
57. L. Rumi, C. Pfleger, P. Spurr, U. Klinkhammer, W. Bannwarth, Org. Process Res. Dev. **13**, 747 (2009)
58. L. Tamborini, P. Conti, A. Pinto, C. De Micheli, Tetrahedron Asymmetry **21**, 222 (2010)
59. T. Jong, A.M. Pérez-López, E.M.V. Johansson, A. Lilienkampf, M. Bradley, Bioconjugate Chem. **26**, 1759 (2015)
60. T.S.A. Heugebaert, B.I. Roman, A. De Blieck, C.V. Stevens, Tetrahedron Lett. **51**, 4189 (2010)
61. M.M.E. Delville, K. Koch, J.C.M. van Hest, F.P.J.T. Rutjes, Org. Biomol. Chem. **13**, 1634 (2015)
62. A. Sniady, M.W. Bedore, T.F. Jamison, Angew. Chem. Int. Ed. **50**, 2155 (2011)
63. C. Wiles, P. Watts, Beilstein J. Org. Chem. **7**, 1360 (2011)
64. L. Lengyel, V. Gyollai, T. Nagy, G. Dorman, P. Terleczky, V. Hada, K. Nogradi, F. Sebok, L. Uerge, F. Darvas, Mol. Divers. **15**, 631 (2011)
65. A.G. O'Brien, Z. Horvath, F. Levesque, J.W. Lee, A. Seidel-Morgenstern, P.H. Seeberger, Angew. Chem. Int. Ed. **51**, 702 (2012)
66. P. Cyr, A.B. Charette, Synlett 1409 (2014)
67. K.A. Roper, H. Lange, A. Polyzos, M.B. Berry, I.R. Baxendale, S.V. Ley, Beilstein J. Org. Chem. **7**, 1648 (2011)
68. K.A. Roper, M.B. Berry, S.V. Ley, Beilstein J. Org. Chem. **9**, 1781 (2013)
69. C.F. Carter, H. Lange, D. Sakai, I.R. Baxendale, S.V. Ley, Chem. Eur. J. **17**, 3398 (2011)
70. J.-C.M. Monbaliu, J. Jorda, B. Chevalier, C.V. Stevens, B. Morvan, Chem. Today **29**, 50 (2011)
71. C.E. Brocklehurst, H. Lehmann, L. La Vecchia, Org. Process Res. Dev. **15**, 1447 (2011)
72. Z. Yu, Y. Lv, C. Yu, Weike Su. Org. Process Res. Dev. **17**, 438 (2013)
73. S.M. Opalka, A.R. Longstreet, D.T. McQuade, Beilstein J. Org. Chem. **7**, 1671 (2011)
74. S.B. Otvos, I.M. Mandity, F. Fulop, ChemSusChem **5**, 266 (2012)
75. G. Kardos, T. Soos, Eur. J. Org. Chem. 4490 (2013)
76. S. Rossi, M. Benaglia, A. Puglisi, C.C. De Filippo, M. Maggini, J. Flow Chem. **5**, 17 (2015)

77. A. Odedra, K. Geyer, T. Gustafsson, R. Gilmour, P.H. Seeberger, Chem. Commun. **2008**, 3025 (2008)
78. T. Fukuyama, M. Kobayashi, M.T. Rahman, N. Kamata, I. Ryu, Org. Lett. **10**, 533 (2008)
79. T. Fukuyama, T. Kawamoto, M. Kobayashi, I. Ryu, Beilstein J. Org. Chem. **9**, 1791 (2013)
80. M. Brasholz, K. von Kanel, C.H. Hornung, S. Saubern, J. Tsanaktsidis, Green Chem. **13**, 1114 (2011)
81. M. Schön, M. Schnürch, M.D. Mihovilovic, Mol. Divers. **15**, 639 (2011)
82. P.Y. Dapsens, C. Mondelli, B.T. Kusema, R. Verel, J. Pérez-Ramírez, Green Chem. **16**, 1176 (2014)
83. C.B. Kelly, C.X. Lee, N.E. Leadbeater, Tetrahedron Lett. **52**, 263 (2011)
84. Y. Wu, W.-Q. Chen, Y.-Q. Zhao, H.-R. Piao, Chin. Chem. Lett. **26**, 334 (2015)
85. C. Wiles, P. Watts, ChemSusChem **5**, 332 (2012)
86. T. Gustafsson, H. Sörensen, F. Ponten, Org. Process Res. Dev. **16**, 925 (2012)
87. S.A.M.W. van den Broek, J.R. Leliveld, R. Becker, M.M.E. Delville, P.J. Nieuwland, K. Koch, F.P.J.T. Rutjes, Org. Process Res. Dev. **16**, 934 (2012)
88. A. Voros, Z. Baan, P. Mizsey, Z. Finta, Org. Process Res. Dev. **16**, 1717 (2012)
89. R.M. Harris, B.I. Andrews, S. Clark, J.W.B. Cooke, J.C.S. Gray, S.Q.Q. Ng, Org. Process Res. Dev. **17**, 1239 (2013)
90. M. Baumann, I.R. Baxendale, Beilstein J. Org. Chem. **9**, 1613 (2013)
91. M. Asadi, S. Bonke, A. Polyzos, D.W. Lupton, ACS Catal. **4**, 2070 (2014)
92. E. Rosatelli, A. Carotti, M. Ceruso, C.T. Supuran, A. Gioiello, Bioorg. Med. Chem. Lett. **24**, 3422 (2014)
93. J.L. Vrijdag, F. Delgado, N. Alonso, W.M. De Borggraeve, N. Perez-Macias, J. Alcazar, Chem. Commun. **50**, 15094 (2014)
94. C. Henry, D. Bolien, B. Ibanescu, S. Bloodworth, D.C. Harrowven, X. Zhang, A. Craven, H.F. Sneddon, R.J. Whitby, Eur. J. Org. Chem. 1491 (2015)
95. M. Brzozowski, J.A. Forni, G.P. Savage, A. Polyzos, Chem. Commun. **51**, 334 (2015)
96. T. Nobuta, G. Xiao, D. Ghislieri, K. Gilmore, P.H. Seeberger, Chem. Commun. **51**, 15133 (2015)
97. T. Doi, H. Otaka, K. Umeda, M. Yoshida, Tetrahedron **71**, 6463 (2015)
98. S.V. Ley, R.J. Ingham, M. O'Brien, D.L. Browne, Beilstein J. Org. Chem. **9**, 1051 (2013)
99. T.W.T. Bristow, A.D. Ray, A. O'Kearney-McMullan, L. Lim, B. McCullough, A. Zammataro, J. Am. Soc. Mass Spectrom. **25**, 1794 (2014)
100. D. Ghislieri, K. Gilmore, P.H. Seeberger, Angew. Chem. Int. Ed. **54**, 678 (2015)

Chapter 11
Outlook

The introduction of the modern assembly line by Ransom Olds in 1903 and its perfection by Henry Ford in 1913 (allowing the production of the Ford Model T in only 93 minutes) had a tremendous effect on today's industry. Ever since, continuous manufacturing is considered undoubtedly as the most suitable approach for large-scale continuous production in every aspect of modern industry (including the fine-chemical and pharmaceutical businesses). With growing environmental and economic demands for the chemical production around the globe, the search for new ways to produce faster and safer, with less labor, but also in a reliable and consistent manner, has become inevitable. The endeavors to meet these new demands have already led to a paradigm change in industry—by moving processes from batch to continuous production. At the same time, the scientific community in the research laboratories is still facing the challenge of the relatively slow uptake of these new technologies that lead away from the round bottom flask towards entirely new synthesis techniques—microwave and continuous flow synthesis. The brilliant work of various research groups around the world has demonstrated the new opportunities that are made now available by the recent advance in technology—intensifying processes by enhanced mass, heat and light transfer; safe manipulation of hazardous materials; and new reaction pathways, all extending the chemical space available for exploration. Nevertheless, the researchers remain somewhat reluctant in exploring this new field. Currently, there is an ongoing discussion, if and in which cases a continuous flow approach should be considered. Although a vast amount of the known chemical reactions can be performed under continuous conditions, the decision has to be taken via a case-by-case approach [1]. As a rather young research field, continuous flow organic synthesis still has to face with various challenges and not all of these can be easily solved. Major problems remain, for example, solid processing, selectivity enhancement, missing unique reactivity, etc. [2]. Another important point is the high cost of commercial flow equipment. Further issue with commercial equipment is to meet the requirements for the envisaged research. Both financial and suitability demands often result in the "do-it-yourself" approach to flow chemistry—building a flow device from available

T. Glasnov, *Continuous-Flow Chemistry in the Research Laboratory*,
DOI 10.1007/978-3-319-32196-7_11

chromatography equipment. This phenomenon is very reminiscent of the development of microwave synthesis, where, in the early days, many modified household microwave ovens have been utilized, making the obtained results mostly irreproducible. Additional considerable hurdle for the implementation of flow processing in the pharma and fine-chemical industry is also the lack of experienced and correspondingly educated personnel combined with the limited ability of flow technology to support diverse synthetic processes. Nonetheless, plant economics leads to the fact that continuous flow processing is mostly reasonable for at a least multiton production of (relatively) simple products.

However, continuous flow synthesis is in its infancy as a research discipline and still has to develop into a fully viable tool for the synthetic chemists, as it was the case with microwave-assisted synthesis at the dawn of the twenty-first century. It is certainly in the hands of the researcher to make the right choice—to use or not to use a continuous flow device for a planned synthesis.

References

1. F.E. Valera, M. Quaranta, A. Moran, J. Blacker, A. Armstrong, J.T. Cabral, D.G. Blackmond, Angew. Chem. Int. Ed. **49**, 2478 (2010)
2. K.S. Elvira, X. Casadevall i Solvas, R.C.R. Wootton, A.J. deMello, Nat. Chem. **5**, 905 (2013)

MIX
Papier aus verantwortungsvollen Quellen
Paper from responsible sources
FSC® C105338

If you have any concerns about our products,
you can contact us on
ProductSafety@springernature.com

In case Publisher is established outside the EU,
the EU authorized representative is:
Springer Nature Customer Service Center GmbH
Europaplatz 3, 69115 Heidelberg, Germany

Printed by Libri Plureos GmbH
in Hamburg, Germany